Politics and Policies for Water Resources Management in India

This comprehensive volume explores the interface between politics and policymaking in the water management sector of India. The authors discuss the nature of the political discourse on water management in India, and what characterises this discourse. They also explore how this discourse has influenced the process of framing water-related policies in India, particularly through the 'academics-bureaucrat-politician' nexus and the growing influence of the civil society groups on policy makers, which are the defining feature of this process, and which have produced certain policy outcomes that are not supported by sufficient scientific evidence.

The book reveals that the social and management sciences, despite being increasingly relevant in contemporary water management, are unable to impress upon traditional, engineer-dominated water administration to seek solutions to complex water problems owing to a lack of interdisciplinary perspective in their research. The authors also examine the current deadlock in undertaking sectoral reforms due to existing water policies not being honoured.

This collection includes several research studies which suggest legal, institutional policy alternatives for addressing the problems in areas such as irrigation, rural and urban water supply, flood control and adaptation to climate variability and change.

It was originally published as a special issue of the *International Journal of Water Resources Development*.

M. Dinesh Kumar is the Executive Director at the Institute for Resource Analysis and Policy in India. He is a leading water resources expert with 28 years of experience undertaking research, action research, consultancy and training in the field of water, agriculture and energy. He has authored six books and six edited volumes on water resources, agriculture and energy, as well as over 170 research articles and papers.

Routledge Special Issues on Water Policy and Governance

Edited by: Cecilia Tortajada (IJWRD) – Institute of Water Policy, Lee Kuan Yew School of Public Policy, NUS, Singapore
James Nickum (WI) – International Water Resources Association, France

Most of the world's water problems, and their solutions, are directly related to policies and governance, both specific to water and in general. Two of the world's leading journals in this area, the *International Journal of Water Resources Development* and *Water International* (the official journal of the International Water Resources Association), contribute to this special issues series, aimed at disseminating new knowledge on the policy and governance of water resources to a very broad and diverse readership all over the world. The series should be of direct interest to all policy makers, professionals and lay readers concerned with obtaining the latest perspectives on addressing the world's many water issues.

Legal Mechanisms for Water Resources in the Third Millennium
Select Papers from the IWRA XIV and XV World Water Congresses
Edited by Marcella Nanni, Stefano Burchi, Ariella D'Andrea and Gabriel Eckstein

Integrated Water Management in Canada
The Experience of Watershed Agencies
Edited by Dan Shrubsole, Dan Walters, Barbara Veale and Bruce Mitchell

Groundwater and Climate Change
Multi-Level Law and Policy Perspectives
Edited by Philippe Cullet and Raya Marina Stephan

OECD Principles on Water Governance
From Policy Standards to Practice
Edited by Aziza Akhmouch, Delphine Clavreul, Sarah Hendry, Sharon Megdal, James Nickum, Francisco Nunes-Correia and Andrew Ross

Urban Resilience to Droughts and Floods
The Role of Policies and Governance
Edited by Cecilia Tortajada, James Horne and Larry Harrington

Politics and Policies for Water Resources Management in India
Edited by M. Dinesh Kumar

For more information about this series, please visit: https://www.routledge.com/series/WATER

Politics and Policies for Water Resources Management in India

Edited by
M. Dinesh Kumar

Routledge
Taylor & Francis Group
LONDON AND NEW YORK

THIRD WORLD CENTRE FOR WATER MANAGEMENT

First published 2020
by Routledge
2 Park Square, Milton Park, Abingdon, Oxon, OX14 4RN

and by Routledge
52 Vanderbilt Avenue, New York, NY 10017

Routledge is an imprint of the Taylor & Francis Group, an informa business

British Library Cataloguing-in-Publication Data
A catalogue record for this book is available from the British Library

ISBN13: 978-0-367-31278-7

Typeset in Myriad Pro
by codeMantra

Publisher's Note
The publisher accepts responsibility for any inconsistencies that may have arisen during the conversion of this book from journal articles to book chapters, namely the inclusion of journal terminology.

Disclaimer
Every effort has been made to contact copyright holders for their permission to reprint material in this book. The publishers would be grateful to hear from any copyright holder who is not here acknowledged and will undertake to rectify any errors or omissions in future editions of this book.

Contents

Citation Information

The chapters in this book were originally published in the *International Journal of Water Resources Development*, volume 34, issue 1 (January 2018). When citing this material, please use the original page numbering for each article, as follows:

Chapter 1

Water management in India: the multiplicity of views and solutions
M. Dinesh Kumar
International Journal of Water Resources Development, volume 34, issue 1 (January 2018)
pp. 1–15

Chapter 2

Water resources development in India
Chandrakant D. Thatte
International Journal of Water Resources Development, volume 34, issue 1 (January 2018)
pp. 16–27

Chapter 3

India's water management debate: is the 'civil society' making it everlasting?
M. Dinesh Kumar and Chetan M. Pandit
International Journal of Water Resources Development, volume 34, issue 1 (January 2018)
pp. 28–41

Chapter 4

Proposing a solution to India's water crisis: 'paradigm shift' or pushing outdated concepts?
M. Dinesh Kumar
International Journal of Water Resources Development, volume 34, issue 1 (January 2018)
pp. 42–50

Chapter 5

Water transfer from irrigation tanks for urban use: can payment for ecosystem services produce efficient outcomes?
L. Venkatachalam and Kulbhushan Balooni
International Journal of Water Resources Development, volume 34, issue 1 (January 2018)
pp. 51–65

Chapter 6

The negative impact of subsidies on the adoption of drip irrigation in India: evidence from Madhya Pradesh
R. P. S. Malik, Mark Giordano and M. S. Rathore
International Journal of Water Resources Development, volume 34, issue 1 (January 2018) pp. 66–77

Chapter 7

Managing water-related risks in the West Bengal Sundarbans: policy alternatives and institutions
Ernesto Sánchez-Triana, Leonard Ortolano and Tapas Paul
International Journal of Water Resources Development, volume 34, issue 1 (January 2018) pp. 78–96

Chapter 8

Techno-institutional models for managing water quality in rural areas: case studies from Andhra Pradesh, India
V. Ratna Reddy
International Journal of Water Resources Development, volume 34, issue 1 (January 2018) pp. 97–115

Chapter 9

Financial performance of India's irrigation sector: a historical analysis
A. Narayanamoorthy
International Journal of Water Resources Development, volume 34, issue 1 (January 2018) pp. 116–131

Chapter 10

Solarizing groundwater irrigation in India: a growing debate
Nitin Bassi
International Journal of Water Resources Development, volume 34, issue 1 (January 2018) pp. 132–145

Notes on Contributors

Kulbhushan Balooni is a Professor of Economics at the Indian Institute of Management Kozhikode, India. He undertakes interdisciplinary research on the environment, resources, development policy and sustainability in the developing countries context, especially those in South and Southeast Asia.

Nitin Bassi is a Senior Researcher at the Institute for Resource Analysis and Policy in New Delhi, India. He has worked on action research projects funded by UNICEF, FAO, EU, ACF, SRTT and SSNNL.

M. Dinesh Kumar is the Executive Director at the Institute for Resource Analysis and Policy in India. He is a leading water resources expert with 28 years of experience undertaking research, action research, consultancy and training in the field of water, agriculture and energy.

Mark Giordano is a Professor of Geography and Cinco Hermanos Chair in Environment and International Affairs in the Edmund A. Walsh School of Foreign Service at Georgetown University, Washington, D.C., USA. His research focuses primarily on the international political dimensions of water and agriculture.

R. P. S. Malik works for the International Water Management Institute in New Delhi, India.

A. Narayanamoorthy is a Senior Professor and Head, Department of Economics and Rural Development at Alagappa University, India. He specialises in the area of irrigation including micro-irrigation, watershed management and farm economics. He was former Member (Official), Commission for Agricultural Costs and Prices, Government of India, New Delhi.

Leonard Ortolano is UPS Foundation Professor of Water Resources and Environmental Planning in the Department of Civil and Environmental Engineering at Stanford University, USA. He is a specialist in environmental and water resources planning, with a focus on the design and implementation of environmental policies and programmes in the United States and developing countries.

Chetan M. Pandit is a former member of the Central Water Commission in Pune, India.

Tapas Paul is the Lead Environmental Specialist in the Environment and Natural Resources Global Practice at the World Bank in New Delhi, India. His World Bank work includes leading work on integrated management of coastal zones in India (2006–2014) and cleaning of the river Ganga (2009–2012).

M. S. Rathore is the Director of the Centre for Economic and Development Studies in Jaipur, India.

V. Ratna Reddy is the Director of the Livelihoods and Natural Resource Management Institute, India. He is an economist who specialises in natural resource management and environmental economics.

Ernesto Sánchez-Triana is the Global Lead for Environmental Health and Pollution Management for the World Bank in Washington, D.C., USA. He has led the preparation of numerous policy-based programmes, investment projects, technical assistance operations and analytical works.

Chandrakant D. Thatte is an Independent Civil Engineering Professional based in Pune, India.

L. Venkatachalam is a Professor at the Madras Institute of Development Studies, India. He teaches and researches in the areas of environmental economics, behavioural economics and economic theory.

Water management in India: the multiplicity of views and solutions

M. Dinesh Kumar

ABSTRACT

There is very limited scientific evidence to support some of the ideas in the water sector that guide India's government policies in these sectors. Further, the interdisciplinary perspective required for the design of economic instruments, institutions, and laws and regulations to implement existing policies is lacking in social scientists. This article discusses the growing debate on water management options for India, the tone and tenor of policy debate and the inconsistency. It summarizes 10 scholarly articles from various authors which reflect the multiplicity of views on water issues and solutions for water management in the country.

Introduction

The issues concerning water, as a basic input for human survival and socio-economic production functions and an environmental good, have always been part of the political discourse in independent India (Iyer, 2011). It is quite imperative that access to and control over water resources is politically contested (Roth, Boelens, & Zwarteveen, 2005). This discourse has included themes as varied as the technologies for harnessing water (large dams vs. small dams), investments in water resources development (public vs. private), the role of communities in water management, the role of the private sector in provision of water-related services, pricing of water and subsidies, and the basic paradigms of water resources development and management. The participation in this political discourse is not merely between political parties and movements, but also between state and Union governments, between government and civil society organizations and groups, between government and academia, and even between academic groups.

This volume is about the interface between politics and policy making in India's water management sector. It discusses the nature of the political discourse on water management in India; what characterizes this discourse; and how this discourse has influenced the process of framing water-related policies in India. Particularly, it focuses on the 'academic–bureaucrat–politician nexus' and the 'growing influence of civil society groups on policy makers', which are the defining feature of this policy making process and which has produced certain policy outcomes that are not supported by sufficient scientific evidence. On the other hand, it also shows that the social and management sciences, in spite of being increasingly relevant in contemporary water management, are unable to convince traditional, engineer-dominated water administrations to seek solutions to complex water problems that they face, owing to lack of interdisciplinary perspective in the research work they produce. It also shows how the lack of seriousness attached to the already existing policies (National and State Water Policies) is creating a deadlock in undertaking sectoral reforms. The volume also includes a few research studies, which suggest legal, institutional policy alternatives for addressing the problems in the water management sector as well as specific areas such as irrigation, rural and urban water supply, flood control, and adaptation to climate variability and change.

Water management options for India: the growing debate

With one of the greatest spatial heterogeneities in hydrological regimes, geo-hydrological environments, climates, and physiographical conditions, as well as socio-ecologic and cultural environments, India offers one of the greatest challenges to water resource managers and policy makers. An illustration of the complex socio-ecology of water in the country is that mean annual rainfall varies from as low as 100 mm in Jaisalmer, western Rajasthan, to 11,700 mm in Chirapunji, Meghalaya. The country's river systems range from one of the most complex and mightiest river systems of the world, the Brahmaputra, to several hundreds of marginalized ephemeral streams which see flows for only a few hours in the whole year. Its groundwater resources range from the richest Gangetic alluvium to some of the lowest-yielding hard-rock aquifers in the plateau. Its human habitations extend from the mountainous sub-Himalayan region to the coastal plains. Its farming systems range from one of the most productive farming systems in the world to one which is as poor as the rain-fed farming systems of eastern Africa. Its agricultural withdrawal of groundwater ranges from 1280 m^3 per capita per annum in water-scarce Punjab to only 130 m^3 per capita per annum in water-rich Bihar. Its urban areas range from the densely populated Mumbai and Kolkatta, with 30,000 persons per square kilometre, to one with only 500 persons per square kilometre. All these make water management decisions extremely complex, not only for the country as a whole, but for regions and sometimes even localities.

A water policy for the country which aims at sustainable, efficient, equitable and harmonious use of water should guide water resource and services managers to take appropriate policy decisions that fit the socio-ecological context of their geographical areas of operation, rather than giving prescriptions for the management action for a given situation. That the policy document should provide a framework for decision making on water management, rather than rigid rules and dogmatic norms, should be accepted as the guiding principle for policy formulation. The policy objectives and the criteria for decision making are important, not the action. To illustrate this point, 'water use' can be taken as an example. Given the wide variation in water supply and demand across the country, the universal policy objective can be economically efficient use of water, with overall water availability and/or the cost of production of water as important criteria for deciding what is efficient use of water for a region or locality, and not water pricing per se. Here, pricing will be an instrument. However, India's water policy debate is largely driven by often diametrically opposed views on policy actions, and is often characterized by ideological positions on several fundamental issues.

Some of the issues being debated are: (1) who should have ownership of and control over natural resources, especially water resources (the nation, the state, or the local community); (2) who can govern and manage water resources (the professionally managed centralized agencies of the states, institutions of local self-governance, or local communities); (3) the ideal scale at which water can be managed (the local level in the village or watershed, the river basin level, the sub-national level, or the national level); (4) which technology should be used to augment water supplies (large water storage and water diversion systems or small water harvesting structures); (5) the rationale of the decision to go for large water projects, i.e. whether it is driven by the need to mitigate water scarcity and floods, or part of electoral politics to win votes, or merely an outcome of the politician–bureaucrat–technocrat–contractor nexus; (6) whether water should be treated as a social good or an economic good; (7) whether water should be treated as a public good or a private good; and (8) how to regulate water use (through top-down state regulations or through social regulation by local communities).

For instance, the fiercest opposition to the privatization of water supply services (in urban and rural areas) in India comes from those who believe that water should be treated both as a social good and a as public good. They believe that privatization of water services would eventually lead to commodification of water, with private utilities operating on a single-point agenda of increasing their profit margins through monopolistic prices, seriously jeopardizing the goal of equitable access and depriving poor communities of the water needed for their basic survival. The suggestions for larger institutional reforms, such as instituting water rights, are also met with similar opposition by these groups, which claim that it would lead to the complete privatization of water. They instead advocate community management of water resources and water services, without ever describing what should be the basis or norm for allocation of water across users within the community, and where the legal sanction for such allocation mechanisms could come from.

On the other hand, the proponents of water supply privatization also have too little to show that it is the only alternative model available to achieve greater affordability of water and improved efficiency of the utilities (Biswas & Tortajada, 2003).

A great deal of opposition to the building of large dams comes from certain civil society groups which view democracy as one where the 'communities' are the ultimate owners of natural resources, and all the power for making decisions regarding their development and use should be vested in them, or in village-level local self-governments (*panchayats*). They perceive the state as an aggressor, which leads to the untenable position that water development technologies, which require involvement of arms of the state for their execution, be it large dams with canals or mega water supply projects, are detrimental because of the human displacement and ecological destruction, which ultimately lead to social conflicts. They say that a major transformation of our thinking about water is needed, as according to them legal, institutional and procedural changes will not work to resolve such conflicts beyond some point without such a transformation (Iyer, 2005).

According to Iyer (2005), that transformation would include an awareness and understanding of water as a scarce and precious resource to be conserved, protected and used with extreme economy; an integral part of nature; a sacred resource; a common-pool resource to be managed by the community or held as a public trust by the state; primarily a life-support substance and only secondarily anything else (economic good, social good, etc.); a fundamental human and animal right; and a bounty of nature to be gratefully and reverentially received and shared with fellow humans (within the state or province or country, or beyond the borders of the country), future generations of humans, and other forms of life.

Some believe that decisions to build large water projects are part of electoral politics to win voters, and are often outcomes of the politician–bureaucrat–technocrat–consultant–contractor nexus (Bosshard, 2004; Iyer, 2005), which breeds corruption. Their other points of contention are the absence of mechanisms to compensate those who are displaced by reservoirs and canals, shoddy rehabilitation carried out in large schemes in the past, and the unscientific criteria followed in projecting the cost of large water projects (Shah & Kumar, 2008). As noted by Kumar and Pandit (2016), the latter are valid concerns, though the recent experience of the multipurpose Sardar Sarovar Project in addressing these concerns has been positive (Jagadeesan & Kumar, 2015).

While corruption is a major concern in all public works in India and needs to be tackled through concerted efforts, the corruption argument is not used against decentralized water conservation projects implemented by village *panchayats* under the Mahatma Gandhi National Rural Employment Guarantee Act, which are often characterized by rampant corruption, nepotism and favouritism, leading to bad planning decisions, poor designs, poor siting of structures and inferior quality of construction. Instead, the operational guidelines for the implementation of the Mahatma Gandhi National Rural Employment Guarantee Scheme were revised, and the new guidelines were supposed to include mechanisms for eliminating the scope of corruption (Shah, 2012), though there was too little change on the ground as a result of this. The belief is that the *panchayats* will always act in the interests of the community, and are relatively corruption-free, and even if it is not so, enough checks and balances could be created at the local level to protect community interests.

On the other hand, the proponents of these models of water resources development, which involve building large reservoirs in water-rich catchments and canal systems to transfer water to water-scarce regions, still do not employ evaluation criteria which are comprehensive enough to capture all costs and benefits (Shah & Kumar, 2008). The result is that resettlement and rehabilitation, whose costs get underestimated in project planning, remain contentious issues in many large projects.

Inconsistent views and incoherent arguments

Inconsistent views and incoherent arguments dominate the water policy debate initiated by civil society groups in India, owing to limited conceptual clarity on many issues. For instance, most of those who consider water resources a common good and stand strongly against privatization of water services promote decentralized water management at the village level, under the pretext that it promotes community participation and thereby equity in distribution of the newly created water source. But what it actually brings is privatization and greater inequity in access to the scarce resource, through elite capture. The reason is that under a decentralized management model, the rural elite are often able to influence site selection, to get maximum personal benefits. It is hardly realized that decentralized management of the resource at the local level and equity in resource distribution are often incompatible.

Further, those who are ardent promoters of access equity in resource use blindly support heavy public subsidies for water supply services in both the domestic and irrigation sectors and energy subsidies in agriculture, under the pretext that such subsidies would help poor communities access water at low cost (Mukherji, Shah, & Banerjee, 2012). But it is well established a large share of the subsidy benefits in irrigation and electricity supply in the farm sector are actually appropriated by large, resource-rich farmers (Vashishtha, 2006a, 2006b). And in the case of water supply, the major proportion of the subsidies is cornered by rich households that consume a lot of water, while the poor who do not even enjoy water connections get hardly get any share of the subsidy benefits (Kumar & Pandit, 2016).

The same subsidy argument is extended to support heavy subsidies for solar irrigation pumps in the Gangetic Basin. Since India is on a path to reduce its carbon emission intensity under international obligations and the carbon footprint of agriculture is on the rise, the positive externality of emission reductions is being used as an excuse for high capital subsidies of solar photovoltaic (PV) systems (Shah, Pradhan, & Rasul, 2016).

Those who are against water-sector privatization support intensive use of groundwater through well irrigation, which is almost entirely in the private domain in India, and individual/community initiatives for recharging aquifers using surface water, which is in a common property resource. More importantly, they are very vocal in their criticism of public investments in irrigation (Shah, 2016), which is resulting in the government gradually losing control over the use of water. While caution towards privatization of water supply service is sensible, the solution lies in greater public investment in irrigation projects, not in excessive use of groundwater through private farmer initiatives.

The debate on whether water is a social good or an economic good is often used to argue for or against pricing of water. The fact that water is an economic good, when used in production functions such as irrigated agriculture, dairy farming and industry, is not suggestive of the need to price the resource. Instead, the marginal returns from the use of water for economic production functions can be a basis for fixing the prices, if there is a significant cost involved in its production and supply for these economic activities, as the marginal returns give an indication of the price users would be able to pay. In other words, when the resource becomes scarce, with high cost of production and supply, only those uses wherein the marginal returns are higher than the unit cost can be sustained. Conversely, if water is available in plenty in the natural system, pricing of water may not be a big concern, even when allocated for productive purposes, if the cost of production and supply of water is insignificant.

On the other hand, the recognition that water becomes a social good when allocated for human consumption doesn't take away the right of the water utilities to charge a price for it, if it is to recover the cost of production and supply, though affordability should be a major concern in fixing the unit price of water. This is because lack of access to a minimum amount of fresh water could severely constrain social advancement and at times even threaten the very survival of the community itself. This essentially means that mechanisms should be devised to offer targeted subsidies to the poor when prices become high, so as to protect the right to life – not free water for all.

Policy advocacy by civil society groups

Often, many virtues are attached to decentralized management of water supply services, especially rural drinking water supply; community management of water resources, as against centralized water management institutions; import of agricultural commodities (virtual water) from water-rich regions to water-scarce regions, as against physical transfer of water; small water harvesting and recharge initiatives against large dams; and rainfed farming against irrigated agriculture (Iyer, 2008; Kulkarni & Shankar, 2009; Shah, 2009; Shankar, Kulkarni, & Krishnan, 2011). These virtues include: decentralized management of drinking water services, to be efficient, equitable and sustainable; community management of resources, for greater basin/aquifer-wide equity in water access, while addressing community interests and concerns through their participation (Kulkarni & Shankar, 2009; Shankar et al., 2011); virtual water trade, to be ecologically sustainable and to avert huge submergence of land and social displacement, which large water-transfer projects involve (Iyer, 2008); small water harvesting and recharge, to be more cost-effective and capable of producing high incremental returns per unit of water generated (Shah, 2009); and rainfed farming, to be ecologically sustainable.

But there is little evidence to support such claims, especially on matters concerning community management of water supplies and decentralized water management. On the contrary, there is enough evidence to prove that some of these radical views are flawed – be it the potential of rainfed farming to feed the country; the potential of rainwater harvesting to augment water supplies for expanding irrigation in arid and semi-arid regions (Glendenning & Vervoort, 2011; Kumar, Turral, Sharma, Amarasinghe, & Singh, 2008); or the promise of trade in virtual water. However, advocacy groups have been highly influential in shaping water- and energy-related policies in India, as is evident in many policies, programmes and projects of the government during the past couple of decades.

Some of the examples are the recent launching of the Neeranchal project for watershed development in rainfed areas by the Ministry of Rural Development, based on the recommendations of the Parthasarathy Committee (Parthasarathy Committee Report, 2006); continuation of the employment guarantee programme involving the rural labour force for digging pits, building unpaved roads and construction of small water bodies (such as ponds and tanks) by the present central government under the Mahatma Gandhi National Rural Employment Guarantee Act, in spite of growing criticism of the massive corruption involved in the working of the scheme (Mukherjee, 2016); large capital subsidies from the central government for solar PV systems for farmers (Shah, Durga, Verma, & Rathod, 2016); the programme for separation of agricultural feeder lines from the domestic ones in some states to reduce power theft by agricultural consumers (Khanna, 2013); and the launching of a scheme for construction of farm ponds in farmers' fields for crop management in Maharashtra (Kale, 2017).

Part of the reason is the acceptance of some of these ideas by the international development agencies and donor bureaucracies, which enjoy great legitimacy in the international arena. Another reason is the enormous clout these advocacy groups enjoy in government and policy circles.

Similarly, some members of civil society used every opportunity to push their own agenda for the water sector while being part of government committees on reform in the water, rural development and agriculture sectors. The latest example is the Mihir Shah Committee report on restructuring of the Central Water Commission and the Central Ground Water Board. Instead of suggesting changes based on a proper diagnosis of the problems facing these sectors and the agencies, facts were often misrepresented. For wider acceptability of their prescriptions, which are based on outdated concepts, the ideas were sold as a 'new paradigm in water management' (Shah, 2016). As protests grew from the senior cadres of one of the agencies (the Central Water Commission), against the recommendations of the report, the media was heavily used to publicize the views of the committee and to put pressure on the governments to buy into their ideas (Economic & Political Weekly, 2016).

The academic–bureaucrat–politician nexus

India's ability to maintain good economic growth and move people out of poverty requires solving several complex problems relating to the supply and demand for water, energy and food. This more often requires hard decisions on the part of the state bureaucracies dealing with water, energy and agriculture, supported by strong political leadership (Kumar, 2014a).

A growing proportion of Indian academics believe that for a democratically elected government to survive, decisions concerning the management and use of resources, which directly affect the livelihoods of millions of ordinary people, should be driven by popular concerns rather than macroeconomic interests. Addressing popular concerns suits the political class better than meeting the macroeconomic interests as it produces immediate electoral gains. These academics also believe that these decisions have to suit the political needs of the elected governments. A related assumption is that ordinary people are only interested in measures that produce immediate gains and are not concerned with long-term economic management.

As Kumar (2014a) notes, to a great extent this is true in India, where a large proportion of the people living in villages and cities are concerned with their immediate survival needs. However, it is also true that a large percentage of the population is quite concerned with long-term economic growth, and their numbers are growing, as the economy is on an upward swing. Good politics also implies that this segment of the population should not simply be ignored for the sake of the rest. Conversely, it is also true that continuing with such populist measures may lead to crippling of the economy, forcing the governments to resort to harsher measures and leading to political backlash.

In good statecraft, it is also the responsibility of the technocrats, the bureaucracy and the partners in policy making, backed by research-based knowledge, to educate the rulers on the need for pragmatic policies that strike a balance between populism and long-term economic management, and to inform them that such an approach is in the interest of the ruling political class. The political class in turn can educate the electorate. This can foster mass support for policy reforms (Kumar, 2014a).

It is the duty of the state to frame policies, rules, administrative structures and laws from time to time to make sure that the subjects have more or less equal opportunities to access common-property resources like water for beneficial uses (Ford Foundation, 2010), while it is the duty of academicians and scholars to help governments design the right kind of policies, legal frameworks, institutions, pricing mechanisms, taxes on development, and allocation and use of these resources based on considerations of productivity, equity, sustainability and social justice. But the state should make sure that the right kind of institutional regime exists to bring checks and balances to ensure quality in academic work that feeds into government policies (Kumar, 2014a).

As Kumar (2014a) notes, some academics advise politicians and bureaucrats that providing free power, fuel subsidies, free power connections and free water would help eradicate poverty (and bring them more votes), and that market-based solutions would harm the interests of the poor. Essentially, by doing this, they are approaching complex economic problems with kid gloves. This academic–bureaucrat–politician nexus is costing the economy hugely. While the argument of inefficient and unsustainable resource use is being used by a few scholars and development thinkers to counter the government policy of the pervasive subsidy (Bhatia & Gulati, 2004; Gulati, 2011), the counterargument, as noted by Kumar (2014a), is that attempts to do away with subsidies in the water and electricity sectors would be politically very risky for ruling parties.

Poor application of social sciences

While the conventional engineering solutions for water management are increasingly inadequate for the growing water problems, the institutions that were built on the foundations of civil engineering and infrastructure building for supply augmentation are becoming less effective. It is increasingly acknowledged that the solutions to the complex problems of water management cannot be found within the disciplines of science and engineering (hydrology, geohydrology, water engineering/technology) alone, and that there is a need to integrate knowledge from social science disciplines – economics, sociology, public administration, and law – with that from the traditional disciplines.

As water allocation and water resources management take precedence over water development under growing scarcity problems (Saleth & Amarasinghe, 2010), this interdisciplinary knowledge is required in the field of water resource economics, environmental economics, water law, organizational behaviour, etc., for valuation of water and design of economic instruments (for water demand management), institutions, laws/regulations and policies (Kumar, Bassi, Venkatachalam, Sivamohan, & Vedantam, 2012). These, combined with the technical system, make for a good water management system. This need is increasingly shown in the acceptance of the concept of Integrated Water Resource Management at the highest levels in the water resources bureaucracy and government policy-making bodies, which call for interdisciplinary approach in water management.

But sadly, the specialists from the social and management sciences are not able to impress the traditional water bureaucracies that are dominated by civil engineers, due to the lack of interdisciplinary perspective in the work of the former. One reason for this is insufficient knowledge of the fundamental, technical issues in water management. Another issue is their limited understanding of water management institutions, born mainly out their experience working with small, micro-level community organizations (such as watershed management committees and water user associations in canal commands with limited operational domains), and lack of experience in crafting modern, high-level institutions that are capable of bringing about behavioural changes in water users, such as water rights systems, water taxes, water resource fees, pollution taxes and water resource regulatory agencies.

National and state water policies: toothless tigers?

Independent India has had three National Water Policies, the first in 1987 (Government of India, 1987, source: https://www.rainwaterharvesting.org/Downloads/nwp1987.pdf), the second in 2002 (Government of India, 2002), and the third in 2012 (Government of India, 2012). All of them highlighted the importance of provision of water for basic survival needs and gave top priority to drinking water (water for human needs) in water allocation decisions. However, these policies did not prescribe, nor were they followed by creation of rules, norms and institutional mechanisms to ensure water allocation for drinking water security. The water policies of the states also give top priority to drinking water (Thatte, in press). However, when it comes to norms and institutional mechanisms for water allocation, nothing really exists on the ground. Though the District Collectors are empowered to earmark water from surface reservoirs (meant for irrigation, etc.) during droughts to make water available for drinking purposes, the existing laws are inadequate to deal with groundwater, as the de facto rights in groundwater are still attached to land ownership.

An important prescription in the water policy documents of 2002 and 2012 is taking the river basin as the unit for planning of water resources, to promote sustainable water resources development. However, so far no serious attempts have been made to implement this idea, and the various agencies concerned with water resources development and management at the state level, in whose domain the related activities fall, continue to act independently, in a sectoral and segmented manner. The only positive recent development is the plan to develop river basin management plans for three of the large river basins in India (Krishna, Godavari and Mahanadi), an initiative of the national Ministry of Water Resources.

Resource assessment and planning of groundwater and surface water resources is carried out separately (by the respective agencies concerned, both at the central level and state levels), without attention to their interconnectedness. Studies have shown that intensive use of groundwater in the upper catchments of river basins results in reduced streamflows, as excessive groundwater draft reduces the lean-season flows (base flow) (Kumar, 2010; Srinivasan et al., 2015).

Similarly, the wings of the Water Resources Department of various state governments, which carry out catchment-wise assessment of surface water potential and plan large and medium-size irrigation/multipurpose water projects, do not take cognizance of the myriad of minor irrigation and watershed development projects that are being planned in the same catchments by other agencies, such as the Minor Irrigation Department and Watershed Development Agencies, and vice versa. Such uncoordinated planning leads to over-appropriation of the resource. Intensive watershed development desilting of tanks, etc., reduces inflows into reservoirs downstream (James et al., 2015). In sum, the policies have not resulted in any organizational restructuring in India's water resource administrations.

The need to treat water as an economic good and the pricing of water to reflect its scarcity value are well recognized in the water policies of 2002 and 2012 (Thatte, in press). Yet, no state government is willing to charge for water supplied by public irrigation schemes on a volumetric basis, and the water access charged by crop area is heavily subsidized, not even covering the full operation and maintenance costs of the schemes (Kumar, 2010), though estimates of the price elasticity of water demand in many developed countries (Australia, North America and Western Europe) show positive values of elasticity (Hassel, 2007; Hoffman, Worthington, & Higgs, 2006). In very few cities and towns in India do the municipalities/corporations or the autonomous water utilities charge for water on a volumetric basis, and not all domestic water connections are metered (Asian Development Bank, 2007; Kumar, 2014b). This gives no incentive for consumers to use water efficiently and reduce waste. In the farm sector, the electricity supplied for groundwater pumping is not charged on a pro rata basis. While some state electricity utilities charge for electricity on the basis of connected load, many states are offering free electricity. In fact only West Bengal successfully introduced metering of electricity in the farm sector for groundwater pumping (way back in 2006), and charges electricity tariffs comparable with the cost of production and supply. What emerges from these examples is that the water policy is never taken seriously by the governments with its bureaucrats and executive wing.

The contents of this volume

Since Independence, government decisions on large-scale public sector investments in irrigation in India have been heavily influenced by academic research, mainly from social science disciplines (agricultural economics, development economics and rural sociology), showing positive impacts of irrigation on food security, crop yields, agricultural productivity, rural poverty and employment (Bhattarai & Narayanamoorthy, 2003; Hussain & Hanjra, 2004). During the Green Revolution (1968–85), India invested heavily in crop research, infrastructure and market development, supported by development of appropriate policies (Pingali, 2012). Support of various public-sector institutions in disciplines as varied as hydrology, water engineering, irrigation science, crop sciences, animal sciences, agricultural economics and rural development from the country's government and international aid agencies, for science and technology development and policy formulation, was a key feature. For a long time, these public-sector institutions worked in a synergetic fashion towards the goals of food self-sufficiency, rural development and poverty reduction. International public goods institutions also contributed significantly to research, especially in the area of plant breeding (Pingali, 2012).

However, much has changed in the way knowledge is generated in the fields of science and technology and public policy. Today, civil society organizations and private-sector institutions also play a significant role in knowledge generation in agricultural sciences, water technology, energy, etc., and also influence government policies with regard to the choice of technologies. There is very little state control of the nature and quality of research that feeds into government policy.

The special issue has a collection of 10 articles, which discuss how the current political economy, characterized by a strong academic–bureaucrat–politician nexus, and the ever-increasing presence of civil society groups, gets played out in almost all domains of water management (technology for water harnessing, water pricing, investments in water resources development, private-sector participation in provision of water supply services, community participation in water management) to influence policies in their respective domains, often through the planting of certain myths. They also show how social scientists are failing India's water administration by not being able to come up with designs for effective legal and policy instruments and institutions for management of water resources that are in line with the country's water policy and use sound principles to address the current challenges. Some articles also illustrate how an integrated understanding of the problems from physical and social science perspectives can help frame meaningful policies and at the same time challenge long-held views about what can work.

Thatte (in press) tracks the history of water resources development in India, particularly how the scientific and technological advancements helped evolve new paradigms in water resources development, beginning in the precolonial era. It also reviews the current legal, institutional and policy regimes governing the development and use of water resources in India, the overall performance of the water sector in relation to meeting water needs for food and agricultural production, human needs and nature, and human resources development in the sector, and identifies some of the key water management challenges for future.

Kumar and Pandit (2016) discuss certain ideas in water resources development and management which have taken root in India's water sector through their success in transforming India's food security and community health situations. In their view, these ideas are still relevant in India's water sector, but are increasingly becoming unpopular among civil society groups. The article systematically investigates the reasons for this unpopularity, and also shows how civil society groups have been able to convince certain academics and policy makers to abandon these ideas and pick up the 'alternatives' that were not always useful and effective.

The article by Kumar (2017) is a critique of the Mihir Shah Committee's report on restructuring the CWC and CGWB. It argues that the recommendations of the committee, which envisage a newly proposed National Water Commission that will carry out water resources development planning at the basin level and implement Integrated Water Resources Management, are not based on any sound understanding of the federal nature of water administration in India, water-sector performance or the problems confronting it. The article argues that the analysis of the performance of surface irrigation systems, used by the committee to criticize the irrigation bureaucracy, is flawed and based on outdated concepts of irrigation efficiency. It further argues that the 'paradigm shift' in water management proposed in the report, which over-emphasizes the role of groundwater in meeting future water needs, reflects the professional bias of its members against large water infrastructure, and wishful thinking about what schemes like aquifer mapping can achieve. In the views of the author, the committee, having not recognized the role of water pricing and water rights and legal reforms in the sector in bringing about behavioural changes, fails to make suggestions for institutional reform that can improve the sector's performance.

Rapid urbanization in India is forcing many of its state governments to transfer irrigation water to urban use. But the economic and environmental consequences of this water allocation, which is happening in a legal and institutional vacuum, are not adequately understood. In Tamil Nadu, transfer of water from irrigation tanks is happening on a large scale. Such water transfers not only harm the economic welfare of innumerable users of water-based ecosystem services in rural areas but also reduce their incentives to manage the water bodies on a sustainable basis. Since the poorer people are largely dependent on these ecosystem services for their livelihood, reduction of ecosystem services due to transfer of water can increase poverty and inequality among water users. Appropriate institutional mechanisms could bring in collective action among different stakeholders and convert the trade-off in welfare across different stakeholders into economic opportunities. Venkatachalam and Balooni (2017) explore the possibility of introducing such institutional mechanisms in the form of 'payment for ecosystem services' at the grass-roots level in the Indian context in return for reallocation of water from irrigation to urban uses so that it can produce non-zero-sum outcomes for villagers, farmers, urban consumers and governments.

In India, drip irrigation coverage has expanded slowly. Studies attribute this to several factors, including physical, socio-economic, institutional and financial (Kumar et al., 2008; Schoengold & Zilberman, 2007). Financial issues are important for smallholders (Malik & Rathore, 2012; Malik et al., this volume). Governments have provided capital subsidies to address the financial issues facing smallholders. Malik, Giordano, and Rathore (2016) argue that rather than improving access to drip systems, the subsidy system holds the farmers back, because the technical requirements for the potential adopters to become eligible, the highly bureaucratic processes of applying for subsidy, and the pricing incentives turn many drip providers into rent-seeking agents rather than service providers to farmers, leading to price increases of 40% or more. The article further argues that if capital costs are truly the constraint on expansion of drip irrigation in India, alternative models to address them are available.

Sánchez-Triana, Ortolano, and Paul (2016) examine alternative policy scenarios for reducing the vulnerability of the people living in West Bengal's Sundarbans, which experience water-related threats – sea level rise, salinization of soil and water, waterlogging, and flooding due to embankment overtopping and failure – along with widespread poverty and anticipated effects of climate change. Of the four alternative policy directions examined, the recommended approach involves embankment realignment and incentives that encourage the gradual, voluntary migration of residents from high-risk parts of the Sundarbans to safer areas offering better employment opportunities. The authors describe how the study findings have been used to inform ongoing deliberations to build consensus on future policy directions to reduce the region's vulnerability to natural disasters by the government of West Bengal.

Reddy (2016) examines the rationale, technologies, economics and institutional modalities in the work of the corporate-supported NGOs, independent NGOs and private firms in the rural drinking water sector to assess their potential in terms of providing lessons for designing policies for sustainable service delivery at scale. The study shows that the economic viability of this service is weak for their current scale of operation. Their services are limited to drinking water, so the water quality problems facing the communities are only partially addressed. The author argues that public-private-community partnerships are equally, if not more, effective in terms of economic viability as well as long-term sustainability. Adopting appropriate technologies could help address the water quality issues in a more comprehensive manner. Based on the experiences, the author makes some suggestions for future policies towards strengthening public-private-community partnership.

Low financial recovery has been a major problem in the irrigation sector and is threatening the sustainability of public irrigation. Some Indian states have initiated bold reforms in the irrigation sector to improve its physical, financial and economic performance, while some have raised water rates to improve financial recovery. But not many detailed studies are available which investigate the issue of why financial recovery in the irrigation sector in India is so poor; whether the reforms introduced by some states improved their cost recovery; and what kind of nexus exists between agricultural development and financial recovery across the states. In his article, Narayanamoorthy (2017) attempts to address some critical questions in irrigation-sector reform, using temporal and spatial data for a period of four decades. The study shows that despite the substantial increase in area under canal irrigation over the years, there has been a consistent decline of cost recovery in irrigation projects. Recovery was better in agriculturally less developed states than in agriculturally developed states. The insignificant relationship between agricultural development and financial recovery rate across states reinforces the fact that farmers' ability to pay for water is not so much dependent on their socio-economic conditions. The experience with irrigation reforms suggests that financial recovery can be improved if revision of water rates is complimented by financial and institutional reforms of the irrigation administration.

Heavy subsidies for electricity and diesel to pump groundwater for irrigated agriculture, combined with lack of regulations on water withdrawal, are resulting in both groundwater over-exploitation and higher carbon emissions (Bassi, 2015). Grid-connected, farm-level solar

PV systems with a 'feed-in tariff' (a payment made to farmers generating their own electricity through methods that do not contribute to the depletion of natural resources or carbon emission, proportional to the amount of electricity generated) are also advocated by a few researchers on the basis of the assumption that they will create an incentive among the farmers to save the electricity so generated in their farms to sell to the utility, thereby also creating an incentive for groundwater conservation (Shah et al., 2016). However, little attention is paid to the comparative economics of solar PV systems and the impact such capital subsidies will have on the public exchequer (Bassi, 2015). The article by Nitin Bassi (2017) examines whether solar pumps for groundwater irrigation are technically feasible and economically viable in India and to what extent they offer incentives for farmers to save groundwater.

Overall, the analyses presented in several of the articles in this special issue show that there is very little scientific evidence to support some of the recent ideas in the water, agriculture, energy and climate sectors that are guiding India's policies with regard to choice of technologies and institutions to solve various problems in these sectors. There is inadequate scrutiny of the information available in the public domain in the form of 'evidence' produced by influential civil society groups and academics in support of these ideas, due to the absence of strong institutional regimes controlling the quality of academic research and the strong academic–bureaucrat–politician nexus. Another major issue is that the interdisciplinary perspective required for the design of economic instruments, institutions, and laws and regulations to implement the existing water policies is lacking among professionals in the social sciences, and as a result the water policies in the country largely remain 'toothless tigers'. The analyses presented in a few articles also suggest legal and institutional policy alternatives to address the problems in the water management sector, as well as specific areas such as irrigation, rural and urban water supply, flood control, and adaptation to climate variability and change.

Disclosure statement

No potential conflict of interest was reported by the author.

References

Asian Development Bank. (2007, November). *2007 Benchmarking and data book of water utilities in India*. A partnership between the Ministry of Urban Development, Government of India and the Asian Development Bank.

Bassi, N. (2015, March 7). Irrigation and energy nexus: Solar pumps are not viable. *Economic and Political Weekly*, 50, 63–66.

Bassi, N. (2017). Solarizing groundwater irrigation in India: A growing debate. *International Journal of Water Resources Development*. doi: 10.1080/07900627.2017.1329137

Bhatia, B., & Gulati, M. (2004). *Reforming the power sector: Controlling electricity thefts and improving revenue, public reforms for the private sector*. Washington, DC: The World Bank.

Bhattarai, M., & Narayanamoorthy, A. (2003). Impact of irrigation on rural poverty in India: An aggregate panel data analysis. *Water Policy, 5*, 443–458.

Biswas, A. K., & Tortajada, C. (2003, October 16). Colombo's water supply: A paradigm for the future? Special feature. *Asian Water*.

Bosshard, P. (2004, April). *The World Bank at 60: A case of institutional Amnesia? A CRITICAL LOOK at the implementation of the bank's infrastructure action plan*. International Rivers Network.

Economic and Political Weekly. (2016, December 24). Water governance. *Special Issue, Economic and Political Weekly, 51*, 19–62.

Ford Foundation. (2010). *Expanding community rights over natural resources: Initiative overview*. New York, NY: Author.

Glendenning, C. J., & Vervoort, W. (2011, February). Hydrological impacts of rainwater harvesting (RWH) in a case study catchment: The Arvari River, Rajasthan, India: Part 2. Catchment-scale impacts. *Agricultural Water Management*, 715–730.

Government of India. (2002, April 1). *National water policy*. New Delhi: Ministry of Water Resources.

Government of India. (2012). *National Water Policy (2012)*. New Delhi: Ministry of Water Resources.

Gulati, M. (2011, April 2). 50% savings, 100% free power. *The Times of India*. Retrieved from https://epaper.timesofindia.com

Hassel, T. (2007). *Promoting behaviour change in household water consumption: A literature review*, prepared for Smart Water. Victoria, Australia.

Hoffman, M., Worthington, A. C., & Higgs, H. (2006). *Urban water demand with fixed volumetric charging in a large municipality: The case of Brisbane*. Brisbane: Faculty of Commerce papers, University of Wollongong.

Hussain, I., & Hanjra, M. A. (2004). Irrigation and poverty alleviation: Review of the empirical evidence. *Irrigation and Drainage, 53*, 1–15. doi:10.1002/ird.114

Iyer, R. R. (2005, October). The politicisation of water. *Info change News & Features*.

Iyer, R. R. (2008). Water: A critique of three basic concepts. *Economic and Political Weekly, 43*, 15–17.

Iyer, R. R. (2011). National water policy: An alternative draft for consideration. *Economic and Political Weekly, 46*, 201–214.

Jagadeesan, S., & Kumar, M. D. (2015). *The Sardar Sarovar project: Assessing economic and social impacts*. Delhi, India: Sage Publications.

James, A J., Dinesh Kumar, M., Batchelor, J., Batchelor, C., Bassi, N., Choudhary, J., ... Priti, K. (2015). *Catchment assessment and planning for watershed management*, Vol. 1. Main Report, World Bank Group.

Kale, E. (2017). Problematic uses and practices of farm ponds in Maharashtra. *Economic and Political Weekly, LII*, 20–22.

Khanna, A. (2013, October 7). Separate power feeders can greatly improve rural electrification. *Down to Earth*.

Kulkarni, H., & Shankar, P. S. V. (2009). Groundwater: Towards an Aquifer management framework. *Economic and Political Weekly, 44*, 13–15.

Kumar, M. D. (2010). *Managing water in river basins*. New Delhi: Oxford University Press.

Kumar, M. D. (2014a). Of statecraft: Managing water, energy and food for long-term national security. In M. D. Kumar, N. Bassi, & A. Narayanamoorthy (Eds.), *Water, energy, food security nexus: Lessons from India for development* (pp. 211–220). London, UK: Routledge/Earthscan.

Kumar, M. D. (2014b). *Thirsty cities: How Indian cities can manage their water needs*. New Delhi: Oxford University Press.

Kumar, M. D. (2017). Proposing a solution to India's water crisis: 'Paradigm shift' or pushing out-dated concepts? *International Journal of Water Resources Development*. doi:10.1080/07900627.2016.1253545

Kumar, M. D., Bassi, N., Venkatachalam, L., Sivamohan, M. V. K., & Vedantam, N. (2012). *Capacity building in water resources sector of India*. (Occasional Paper # 4). Hyderabad: Institute for Resource Analysis and Policy.

Kumar, M. D., & Pandit, C. (2016). India's water management debate: Is the civil society making it everlasting? *International Journal of Water Resources Development*. doi:10.1080/07900627.2016.1204536

Kumar, M. D., Turral, H., Sharma, B. R., Amarasinghe, U., & Singh, O. P. (2008). Water saving and yield enhancing micro irrigation technologies in India: When do they become best bet technologies? In M. D. Kumar (Ed.), *Managing water in the face of growing scarcity, inequity and declining returns: Exploring fresh approaches*. Vol. 1, proceedings of the 7th Annual Partners' Meet of IWMI-Tata Water Policy Research program, ICRISAT, Hyderabad.

Malik, R. P. S., Giordano, M., & Rathore, M. S. (2016). The negative impact of subsidies on the adoption of drip irrigation in India: Evidence from Madhya Pradesh. *International Journal of Water Resources Development*. doi:10.1080/07900627.2016.1238341

Malik, R. P. S., & Rathore, M. S. (2012). *Accelerating adoption of drip irrigation in Madhya Pradesh*. New Delhi: AgWater Solutions Project Case Study, International Water Management Institute.

Mukherjee, S. (2016, February 2). 10 years of MGNREGA: How the Modi government was forced to adopt the scheme, *Business Standard*.

Mukherji, A., Shah, T., & Banerjee, P. (2012). Kick-starting a second green revolution in Bengal. *Commentary, Economic and Political Weekly, 47*, 27–30.

Narayanamoorthy, A. (2017). Financial performance of India's irrigation sector: A historical analysis. *International Journal of Water Resources Development*. doi:10.1080/07900627.2017.1298998

Parthasarathy Committee Report. (2006). *From Hariyali to Neeranchal: Report of the technical committee on watershed programmes in India*. New Delhi: Department of Land Resources, Ministry of Rural Development, Government of India.

Pingali, P. L. (2012). Green revolution: Impacts, limits, and the path ahead. *Proceedings of the National Academy of Sciences, 109*, 12302–12308.

Reddy, V. R. (2016). Techno-institutional models for managing water quality in rural areas: Case studies from Andhra Pradesh, India. *International Journal of Water Resources Development*. doi:10.1080/07900627.2016.1218755

Roth, D., Boelens, R., & Zwarteveen, M. (2005). *Liquid relations: Contested water rights and legal complexity*. New York, NY: Rutgers University Press.

Saleth, R. M., & Amarasinghe, U. (2010). Promoting irrigation demand management in India: Options. *Water Policy, 12*, 832–850.

Sánchez-Triana, E., Ortolano, L., & Paul, T. (2016). Managing water-related risks in the West Bengal Sundarbans: Policy alternatives and institutions. *International Journal of Water Resources Development*. doi:10.1080/07900627.2016.1202099

Schoengold, K., & Zilberman, D. (2007). *The economics of water, irrigation and development*. In R. Evenson & P. Pingali (Eds.), *Handbook of agricultural economics* (Vol. 3, pp. 2933–2977). North Holland: Elsevier.

Shah, T. (2009). *Taming the Anarchy: Groundwater Governance in Asia*. Washington, DC: Resources for the Future.

Shah, M. (2012). Report of the Committee for Revision of MNREGA Guidelines, Submitted by Dr Mihir Shah, Chairperson, Committee for Revision of MGNREGA Operational Guidelines to Ministry of Rural Development. New Delhi: Government of India.

Shah, M. (2016). *A 21st century institutional architecture for water reforms in India*. New Delhi: Final Report submitted to the Ministry of Water Resources, River Development & Ganga Rejuvenation, Government of India.

Shah, Z., & Kumar, M. D. (2008). In the midst of the large dam controversy: Objectives, criteria for assessing large water storages in the developing world. *Water Resources Management, 22*, 1799–1824. doi:10.1007/s11269-008-9254-8

Shah, T., Durga, N., Verma, S., & Rathod, R. (2016). *Solar power as remunerative crop, water policy research highlight # 10*. Anand, Gujarat, India: IWMI-Tata Water Policy Program.

Shah, T., Pradhan, P., & Rasul, G. (2016). Water challenges of the Ganga Basin: An Agenda for accelerated reform. In L. Bharati, B. R. Sharma, & V. Smakhtin (Eds.), *The Ganges Basin: Status and Challenges in Water, Environment and Livelihoods* (pp. 304–320). London: Earthscan/Routledge.

Shankar, P. S. V., Kulkarni, H., & Krishnan, S. (2011). India's groundwater challenge and the way forward. *Economic and Political Weekly, 46*, 37–45.

Srinivasan, V., Thompson, S., Madhyastha, K., Penny, G., Jeremiah, K., & Lele, S. (2015). Why is the Arkavathy River drying? A multiple-hypothesis approach in a data-scarce region. *Hydrology and Earth System Sciences, 19*, 1905–1917. doi:10.5194/hess-19-1905-2015

Thatte, C. D. (in press). Water resources development in India: Where are we headed? *International Journal of Water Resources Development*.

Vashishtha, P. S. (2006a). *Input subsidy in Andhra Pradesh agriculture*. Unpublished manuscript.

Vashishtha, P. S. (2006b). *Input subsidy in Punjab agriculture*. Unpublished manuscript.

Venkatachalam, L., & Balooni, K. (2017). Water transfer from irrigation tanks for urban use: Can payment for ecosystem services produce efficient outcomes? *International Journal of Water Resources Development*. doi: 10.1080/07900627.2017.1342610

Water resources development in India

Chandrakant D. Thatte

ABSTRACT
India, an ancient rural and agricultural society that is rapidly modernizing, receives a fair share of its yearly precipitation in only a few days of the monsoon, with high inter-annual variability. In most of its regions, therefore, India needs to store a large proportion of its annual runoff in reservoirs for use in non-monsoon months. In spite of this strategy being in operation for the last 60 years, India's per capita reservoir storage is relatively small, and water-use efficiency also remains low. Though the overall performance of the water sector in terms of matching of supply and demand has improved, the country remains challenged by deficiencies in laws, regulation policies and institutions, and weakened by a suboptimal work culture in politics, legislature, technocracy and non-governmental organizations (NGOs).

Introduction

In a tropical, ancient, agrarian country like India, various methods were known for using surface water from streams, rivers, natural lakes and man-made ponds spread over its great extent, and groundwater from open wells, for local agriculture and domestic use. The ancient literature is replete with anxious enquiries by rulers, even during the monsoon, about the availability of water in storage reservoirs to manage a possible impending drought. Some storage systems were built to reduce inundation of downstream areas. Policies for water resources development (WRD) evolved to keep up with the increasing needs of the growing population. Irrigated agriculture started with flood canals. However, taming of natural disasters such as droughts and floods remained the prime concern. Even today, the Indian countryside is dotted with some very innovative infrastructure built during those times. Colonial rule in India brought modern thinking and technology for WRD, enabling the construction of some classic structures for irrigation, drinking water supply for cities and generation of hydropower, including spectacular inter-basin water transfer schemes, with the help of scores of erstwhile princely states. WRD schemes serving multiple purposes were preferred wherever feasible, as they proved more flexible and economical.

Beginning with the late 18th century, the world saw enormous growth of interest in science, leading to revolutionary technological changes. The first Industrial Revolution in Britain (1775–1800) heralded the steam engine, the birth of the modern factory and mechanization of the textile industry. The second industrial revolution, in the early 20th century, brought the moving assembly line and mass production. Two devastating world wars followed in the first half of the 20th century, followed by a Cold War. The United Nations was constituted to move towards a world government and bring about global peace. The widely known article 'The tragedy of the commons' (Hardin, 1968) highlighted the problem of misuse of shared resources in the economic development of the new societies. The third industrial revolution was related to applied electronics and information technology. And the coming fourth, enunciated at the World Economic Forum in 2015, aims at the fusion of physical, digital and biological technologies to address governance through new techniques such as genome editing and new forms of machine intelligence.

The technological advancements brought about by these industrial revolutions significantly opened up new opportunities for WRD and management, while throwing up new challenges.

The need for water resources development

India receives about 1170 mm of rainfall annually, a little more than the world average, but with great spatial variation, ranging from less than 200 mm in the Jaisalmer district of western Rajasthan to more than 11,000 mm in the Cherapunji district of Meghalaya in the north-east. The summer monsoon is the main contributor of rain in the country. The winter brings rain or snowfall in the Himalayas and some parts of north India. The monsoon spans four months and displays large interannual variability. Its rainfall is due to: (i) the difference in surface temperatures of the Northern Oceans, the Pacific and Atlantic; (ii) the weather systems of El Niño (synonymous with droughts) and La Niña (synonymous with floods); (iii) Indian Ocean surface temperatures and the atmospheric pressure system on the surface of the eastern Asian oceans; (iv) the Indian Ocean Dipole, that is, the surface temperature difference between the Arabian Sea and the Bay of Bengal; (v) land surface temperatures in north-west Europe; and (vi) and the warm water quantum in the Pacific Ocean.

Given the high inter-annual variability in water availability in each river basin, water resource potential is computed at a certain level of dependability (say, 75% or 60%, depending upon intra-year variability). Water storage in large reservoirs in early times was generally built in response to recurring drought and/or flood events. But such reservoirs were soon given more responsibilities, such as relieving shortages in food, energy or drinking water, as the population grew.

The water resources of a region are developed to respond to the basic hydrologic parameters of a river basin shared in a country or between countries; within or between states of a nation; and within or between river basins. To match judiciously supply with demand, on a sustainable basis and with equity, infrastructure is needed. India is drained by 12 major river systems (and many smaller ones) that are shared by more than one state. About 86% of the water resources already developed and appropriated through various types of infrastructure are presently used for agriculture (irrigated or rainfed), 9% for domestic use and 5% for industry. The patterns of water use in the country are changing fast with urbanization and industrialization, with large proportions of water getting diverted to these sectors, while rapid economic progress keeps raising water demand for agricultural production, as food self-sufficiency remains a core goal of WRD for the country.

Regarding extreme events, droughts continuing over more than a few years affect a major proportion of the country's geographical area every 15–20 years on average; the last episode occurred in 2014–16. Monsoon deficit has impacts that take years to recover from. Indian states presently declare droughts on the basis of percentage of crop production expected at the end of the monsoon for the next agricultural year. Some states, where different crops are grown in two seasons, adopt separate declarations for the *kharif* (rainy season) and the *rabi* (dry season). Drought declaration is followed by measures to deal with: (i) loss in land revenue; (ii) escalating energy bills for groundwater pumping; (iii) recovery of agricultural loans/interest; (iv) crop insurance for the future; (v) electricity connections for agro wells; (vi) drinking water supply through tankers; and (vii) restructuring of farm loans. In addition, relief works to employ the farm labourers, both landed and landless, who are unemployed as a result of suspended agricultural activity, plus organization of fodder camps to feed distressed cattle, become top priority.

Over the decades, the Indian water bureaucracies geared themselves up to deal with droughts and floods. In pre-Independence India, droughts often escalated into famines due to socio-economic and infrastructural deficits. Thanks to years of significant WRD work to assure irrigation and drinking water supply, famines no longer occur, though India's population is now nearly four times that of the early 20th century. Much was done to protect society from the socio-economic fallout of droughts, such as shortage of drinking water, fall in agricultural production, loss of agriculture-related employment, migration, and slower gross domestic product (GDP) growth. Drought-afflicted areas are spread over 17 states of India. Of the 471 districts in the country, 222 are afflicted. Of the geographical area of 329 M ha, 121 M ha is drought-affected. The Drought-Prone Area Programme covers 75 M ha (see https://dolr.nic.in/dolr/dpap.asp), and the Desert Development Programme covers the other 46 M ha (see https://dolr.nic.in/dolr/ddp.asp). The chronic flood-prone area is about one-eighth of the country's geographical area. About 40% of this is possibly protected so far. Encroachment on riverine areas due to urbanization and industrialization, with slum rehabilitation, increases the areas vulnerable to floods. The construction of large dams has played a great role in improved flood management.

India's water policies

India's first National Water Policy (NWP), adopted in 1987, was backed by 40 years of experience in building and operation of multipurpose storage and diversion projects, mainly concerned with irrigation, which helped counter the extremes of droughts and floods, bringing about food self-sufficiency (Government of India, 1987); revisions were adopted in 2002 and 2012 (Government of India, 2002, 2012). Since India's independence, successive Five-Year Plans focused on food self-sufficiency. By the end of the 10th Five-Year Plan (Planning Commission, 2002) food sufficiency was realized with buffer stock. Related laws have been enacted, but none has yet dealt with aspects of WRD which were not present in the first national water policy, like adaptation to climate change; enhancing water availability in view of difficulties in rehabilitation and resettlement, land acquisition, and environmental concerns; demand management and water pricing; and river corridors. However, policy aspects have evolved with experience, e.g. changes in flood and drought management, water supply and sanitation, transboundary rivers, project planning, institutions, and databases, which are basic to any good policy for water management.

The National Water Policy of 1987 voiced the need for a master plan. The 2002 National Water Policy recognized the need for river basin authorities (RBAs), but left enunciation of their scope and powers with the states themselves. Guidelines for setting up RBAs were included in the *Guidelines for Preparation of River Basin Master Plan*, published by the Central Water Commission (2007). However, various other forms of RBA continue to operate in India without change, e.g. the Damodar Valley Corporation, the Bhakra-Beas Management Board and the Narmada Control Authority (NCA), and they do constitute a fairly successful starting point in managing rivers at the basin level. Some river boards have been set up under different laws for different purposes: the Upper Yamuna Canal Board, the Ganga River Basin Authority, and the Tungabhadra and Beas River Boards. They all need upgrading to RBAs. The tribunals set up under the River Water Disputes Act 1958 have created institutions similar to RBAs for the Godavari, Krishna, Narmada and Cauveri basins, which are effective. Except for the NCA, none comes close to the prescriptions for an RBA. The NWP of 2012, in the meantime, advocated state water resources regulatory authorities (WRRAs) along the lines of the successful Maharashtra WRRA, which under judicial gaze successfully realized equitable sharing of basin waters between up- and downstream users during the 2014–16 drought, the worst in the last century.

Rights and laws related to water resources

Ancient India had laws on irrigation consistent with the major user of that time: agriculture. The legal terms presently used are akin to terms mentioned in the *Arthashastra* and *Manusmriti* (200–300 BCE) and the Rigveda and Atharvaveda. From the 16th century, the new rulers married local laws with their own. The British followed their own practices, supporting revenue needs. The Northern India Canal and Drainage Act 1873 (see https://punjabrevenue.nic.in/canal_drain_act1.htm) empowered the Canal Officer. The Easements Act 1882 (see https://www.theindianlawyer.in/statutesnbareacts/acts/i19.html) covered only groundwater, recognizing the right of the landowner to use the water underlying his piece of land, making it an individual negative right, and legitimizing rights through two rules: long use, or prescription and local customs. There followed, under British rule, the Land Acquisition Acts 1894 and 1908 (see https://dolr.nic.in/dolr/LandAcquisitionAct1894.asp); the MP [Madhya Pradesh] Irrigation Act 1931 (see https://www.bareactslive.com/MP/MP335.HTM); the Government of India Act 1935 (see https://lawmin.nic.in/legislative/textofcentralacts/GOI%20act%201935.pdf), in which provinces were given prominence; and the Madhya Pradesh Regulation of Waters Act 1949 (see https://www.lawsofindia.org/pdf/madhya_pradesh/1949/1949MP37.pdf).

In the Indian law framework of 1947, various principles, rules and acts coexisted. The Irrigation Act continued to be the most used and developed water law of the colonial government, and laws that followed after 1947 carried on the promotion of multipurpose storage-based irrigation works. Although the Constitution of India was adopted in 1950, water continued to be governed by the states as per the 1935 Act (Government of India, 2015). The Bombay, Madras, Punjab and Calcutta Provincial Acts continued the legacy, till the states were reorganized in 1958 on a linguistic basis under the Society Registration Act, and the new states produced fresh acts, replacing the colonial ones. Special boards and commissions were set up under these new acts: Irrigation Commission Reports of 1903 (East India (Irrigation), 1903) and 1972 (Government of India, 1972); 'Report of the National Commission on Integrated Water Resource Development: A Plan of Action' (Government of India, 1999); Brahmaputra Board Act (see https://www.wrmin.nic.in/writereaddata/Publications/

anu1732289380.pdf), presently under revision; Water (Prevention and Control of Pollution) Act, 1974 (see https://www.envfor.nic.in/legis/water/wat1.html); Inter-State River Water Dispute Act, 1956 (Government of India, 1956); River Boards Act, 1956 (*The Gazette of India*, 1956), which is under revision; National Rehabilitation and Resettlement Policy, 1999 (under revision after combining with Land Acquisition) (Ministry of Rural Development, 2007); and National Dam Safety Bill (new, awaiting clearance) (Sharma, 2017).

The National Water Policy of 2012 identified for the first time the need for an umbrella Water Framework Law in face of the plurality of laws in India's federal set-up. Central legal provisions contradict state provisions in some ways. For example, Section 7(g) of the Easements Act 1882 states that every landowner has the right to 'collect and dispose' of all water under the land within his property, and all water on its surface that does not run in a defined channel. This is not compatible with a legal framework based on the human right to water which requires water to be allocated it to all – whether landowners or not – on an equal basis, for drinking/domestic use.

Also, during the last three or four decades, the need for water-sector reforms with the following characteristics were identified: (1) water as both a social and an economic good, not either/or; (2) decentralization and participation in integrated water resources development and management; (3) water user associations for all users, irrigation management transfer and redefining the government's role; (4) formation of state WRRAs and award of water entitlements, sustainable development, equity, tariff, transfer and trade; and (5) policies and practices such as pro-poor and pro-nature reforms; private-sector investment (not privatization); (6) water user associations and resource wardens as regulators and facilitators; privatization, corporatization and cooperatives; (7) foreign direct investment and international donors; conflict management; (8) matching needs and availability; (9) restructuring of the agriculture and energy sectors; (10) socially regulated properties with non-tradable rights to water and land; and (11) restructuring to protect farms and farming, the lives and livelihoods of farmers, and the landless as farmhands.

Fragmentation of the water sector in government

Functions relating to WRD – such as irrigation, navigation, power generation, rural water supply, industrial water supply and municipal water supply – are dealt with by several wings of the government, each following its own policy. Planning and implementation of WRD projects should be unitary, but it remains fragmented. A national regulatory agency is necessary to meet judiciously, equitably, collectively and optimally the needs or demands of different water-use sectors. This step was taken in the 1980s with the creation of the national Ministry of Water Resources (which had various other names over the decades symbolizing the linking of its responsibilities to agriculture, power etc., as demanded by the social needs of the times). However, the desired unitary and regulatory function has not yet materialized. Adding to the confusion, the name was recently changed again (presumably to address popular concerns) to the Ministry of Water Resources, River Development and Ganga Rejuvenation. 'River development' here means linking rivers and balancing the flow, but rejuvenation of the Ganga is connected only with the management of water quality. Although not intended, such changes incidentally bring the functions of other ministries under the new title, further adding to the confusion.

The ministry is presently responsible for overall planning, policy formulation, coordination and guidance. It is also responsible for regulation and development of interstate rivers, implementation of tribunal awards, water quality assessment, bilateral and external assistance and cooperation programmes in water resources, and matters relating to rivers common to India and neighbouring countries.

Given the issues arising from the fragmented structure of governance of water resources, even before the first NWP of 1987, several reports by commissions, and statements in parliament about WRD, drew on worldwide experiences and indicated acceptance of concepts and criteria such as the river basin as a planning unit; maximizing net benefits through sizing of WRD projects; and keeping the benefit–cost ratio above 1.5. But interstate WRD conflicts grew, even as basin-wise plans under the National Perspective Plan for long-distance inter-basin water transfer were considered. Earlier, in 1983, a national apex body, the National Water Resources Council (see https://wrmin.nic.in/forms/list.aspx?lid=277) was constituted, with the prime minister as chair and state chief ministers as members, to lay down and periodically review the national water policy; advise governments on procedures, administrative arrangements and regulations for distribution and use of water resources for different users to achieve optimum development and maximum benefits; consider and review development plans framed by the National Water Development Agency and river basin commissions; advise on modalities to resolve interstate differences in implementing such projects; and foster expeditious, environmentally sound and economical WRD. The National Water Resources Council's executive arm, headed by the secretary of the Ministry of Water Resources and with state secretaries as members, was also set up as the National Water Board to prepare an action plan.

Water resources development status, performance and prospects

On average, India annually receives about 4000 billion cubic meters (BCM) of water in the form of rainfall. Easterlies bring winter snow and rain to the Himalayas, feeding the Indus, Ganga and Brahmaputra river systems through monsoon runoff and summer snowmelt. The third system bringing significant precipitation occurs along the storm track of cyclonic depressions (originating more from the Bay of Bengal and less from the Arabian Sea). This contribution is much smaller, less regular and uncertain, yet in some river basins (like the Narmada) can convert a dry year into a wet one. The fourth source is runoff flowing into Indian river systems from adjoining countries. About 400 BCM of such inflow brings the total water resources availability in India to 4400 BCM. Even the severe drought year of 2014, with rainfall of 104 cm (12% below average), yielded about 3400 BCM. Of the 4000 BCM of rainfall, Indian rivers drain over 1900 BCM, which is its average water resources potential. Due to losses, about 1100 BCM (700 BCM of surface water and 400 BCM of groundwater) are usable. About 450 BCM of surface water and 250 BCM of groundwater are presently withdrawn from that 1100 BCM; a large component of the renewable groundwater used comes from return flows from areas irrigated by gravity from surface irrigation systems. The balance potential of 250 BCM of surface water remains untapped due to inadequate funding, environmental activism, interstate disputes, rehabilitation and resettlement issues, and local disagreements. The currently identified inter-basin water transfer (IBWT) programmes, if and when implemented, could add about 200 BCM, in addition to about 100 BCM that could be saved by improving water-use efficiency (WUE), to meet the ultimate needs of 1400 BCM of water resources for the estimated population of 1.6 billion in 2060. Thus the ultimate water resources need has to be met with through (i) entire use of the available 1100 BCM, (ii) IBWT of 200 BCM, and (iii) use of saving of 100 BCM by increasing WUE.

River basin management

Between the Himalayas and the peninsular basins lies a large swathe of extensive alluvium floodplains. The river basins of the north are covered by them, and have fewer viable sites for building large storages than peninsular India, making conveyance systems more expensive. The aquifers of the alluvial plains also hold very large potential, owing to high storage coefficient and transmissivity. Groundwater is abstracted from deep or shallow tube wells, with higher pumping capacity than in peninsular India, serving comparatively larger areas.

Regarding river basin management at the national level, river water dispute tribunals were set up in response to complaints by aggrieved states, and awards have been made by the respective tribunals ordering sharing of surface water resources of the Godavari, Krishna (Tribunals I and II), Narmada and Cauveri basins amongst concerned states. The Cauveri award is under review by the Supreme Court to adjudicate certain legal issues. The Narmada Water Disputes Tribunal was a unique case, wherein the party states admitted a non-basin state (Rajasthan) as a party to the dispute and it was allotted a share, as it did not have another source of water to meet the needs of its population in the South-Western region, bordering the Thar Desert. In other river basins, there are standing agreements between party states. The groundwater resources of these interstate basins have not been considered for allocation by river water dispute tribunals, as they are considered 'available' to the related states (Kumar, 2014).

Internationally, the surface water resources of the Indus basin are developed and regulated under the Indus Waters Treaty of 1960 between India and Pakistan (Biswas, 1992). Barring a couple of disputes that are being adjudicated by arbitrators, overall satisfaction with operation of the treaty continues to be high. The Indian states in the Indus basin also share India's quota, under agreements that are working satisfactorily. The Indus is one of the very few river basins globally with successful integrated water resources management. The Indian Bhakra-Beas Management Board regulates water resources for the Indus system.

The second international basin area lies mostly in the alluvial plains of North India, comprising the three sub-basins of Ganga, Brahmaputra and Meghna. They have a multinational spread, but water sharing and water resources management issues are dealt with separately. For example, Ganga water agreements with Nepal, the upstream riparian (Ministry of Water Resources, River Development and Ganga Rejuvenation 2016), and Bangladesh, the downstream riparian (see https://www.mtholyoke.edu/~ahmad20 m/politics/treaty96text.html), are in force, with agreed monitoring mechanisms. The water quality of the Ganga is looked after by the Ganga River Basin Authority under the Ministry of Water Resources, by the Ministry of Environment and Forests, Central Pollution Control Board, State Pollution Control Boards and concerned state governments. There are at least two action plans in operation in the basin for improvement of river water quality, one for the Ganga and the other for the Yamuna. River boards set up under specific acts are also in operation.

Addressing the water needs of different use sectors

On the surface water front, the strategy involves augmentation of existing storage where required, and possible for use through canals, pipelines etc. to the places of demand. On the groundwater front, the strategy has been to construct open or tube wells and adopt artificial recharge measures where necessary; to adopt conjunctive use of surface water and groundwater; and to implement and complete on priority all intra- and inter-basin water transfer plans.

Water for people

Rural water supply coverage in India has reached 80%. Urban water supply coverage is 88%. Both ought to reach 100% soon. But further improvements are necessary. These include dealing with open defecation, presently practised by 50% of the population, replacing it with dry sanitation and WC service by 2035; increasing wastewater collection and treatment coverage from 25% to 100% by 2025, thereby reducing the water supply requirement for secondary use; compensating for the loss of old ponds and increased runoff in urban areas with storm water drains; and raising hydropower development from 22% to 100% of potential.

Other strategies to address the problem, particularly on the water supply front, include the following: (1) make additional good-quality water available for drinking in deficit areas, and also accommodate future migration of people to urban and industrial sectors; (2) build water-supply schemes from reliable sources of more than 90% dependability, plus dams and infiltration wells for urban areas; (3) prepare thoroughly for droughts and other natural disasters; and (4) carry out long-range planning and implementation of water-supply schemes for urban and industrial areas, including advanced assessment of change in habitat status.

India needs to plan for larger-capacity wastewater treatment systems given the larger volume of wastewater from higher per capita water consumption in the domestic sector for washing and secondary uses, in addition to peri-urban irrigated agriculture. There should also be direct regulation of water supplies to reduce waste. Use of old sources (wells, tanks and ponds), but with improvement in the quality of water during droughts, needs to be adopted wherever possible. Given the highly dispersed population, decentralized drinking water supply sources and wastewater treatment plants have to be planned. The industry must undertake to achieve zero-effluent status in about 20 years. We need to ensure that wastewater used for irrigation is given adequate treatment to avert health risks in the food chain (Amerasinghe, Bhardwaj, Scott, Jella, & Marshall, 2013).

Water for food

The per capita land availability in India was 0.9 ha 50 years ago; going by the population growth trends, it might shrink to about 0.14 ha by 2050. India's geographical area is 329 Mha, of which 47% (142 Mha) is cultivated, 23% is forested, 7% is under non-agricultural use and 23% is wasteland. Land reform is of great importance. Of the cultivated area, 37% is irrigated, producing 55% of India's food; 63% is rain-fed, producing 45% of the food. By 2050, the ratio of irrigated to rainfed land might rise to about 50:50, producing 75:25 of the 500 Mt of food required at that time. As the land under crops might remain the same, land productivity will have to be raised significantly to meet the growing production and consumption needs. Indians largely rely on vegetarian food, but with economic advancement, consumption of non-vegetarian food, with its much larger water footprint, will rise. The demand for feed/fodder will also rise due to the rapid rise of per capita consumption of dairy products. The greater production of meat and dairy products will require more water. It will be possible to meet the future water requirements for agricultural production by creating additional irrigation potential in the country through new schemes, including inter-basin water transfer schemes, and by increasing water-use efficiency in irrigated production. The challenges in the sector include maintaining food and cash crop proportions to match water availability, while meeting domestic food security needs and the microeconomic needs of the producers.

Funding for the WRD, particularly related to irrigation schemes, which shrank in the last three Five-Year Plans, needs to be increased. The Bharat Nirman Scheme for the augmentation of irrigation potential has been instrumental in adding about 15 Mha of potential since 2006. The Accelerated Irrigation Benefits Programme (see https://www.wrmin.nic.in/forms/list.aspx?lid=399), with funding for completion of stalled WRD projects, planned to add 15.4 Mha of irrigation potential. Since 2005, about 8 Mha have been added. The new government has recently provided more funds to increase irrigation potential by 2.8 Mha through the newly conceived initiative Pradhan Mantri Krishi Sinchayee Yojana (Prime Minister Agriculture Irrigation Scheme) (see https://pmksy.gov.in/AboutPMKSY.aspx).

The socio-economic agenda for the future with regard to the water sector should be to deal with hunger, food deficit, under-nourishment, unemployment, health issues and farmer income; and to achieve agricultural gross domestic product (GDP) growth of 4% in the next five years, while ensuring wide dispersal of related information on present status.

Water for nature

Among natural resources, land and minerals are considered 'positive' common resources to be owned and used by the community. Water and air are 'negative' commons, which cannot be owned (carry no property rights), barring usufruct rights.

In meeting water needs for food, agriculture, drinking and domestic use by humans, the quality of water resources available for nature gets adversely affected; some quantity is lost to evaporation, so quality declines. The activists opposing the WRD for food production and human consumption call water appropriation from rivers 'the death of rivers', imputing to rivers the status of a living organism, and discounting their role as providers of natural resources. Indian rivers are mostly ephemeral, and the perennial ones have large seasonal variations in flow. When activists call for restoring rivers to 'natural' status, it is not clear what they mean. River water diverted and used for food production and human needs in excess of what is optimally needed enters into shallow aquifers or returns to the stream. Often, 'dilution' of such return flow is sought to improve water quality as an 'environmental need'. The maxim of 'dilution is not a solution to pollution, but treatment is' is forgotten. Sadly, the minimum flow required to maintain seasonally existing aquatic species has not been worked out.

The surroundings of large surface reservoirs and canals can become new habitats for wildlife, exotic flora and fauna, even in the hot and dry tropics, as seen in the Thar Desert of Rajasthan, which was transformed by the Indira Gandhi Canal. There are similar instances around the world. Environmental activists ignore the fact that relief of economic deprivation is made possible by socio-economic development through proper development of water resources. Without this, deprivation would continue to cause impoverishment and migration of landless or workless farmers and farm labourers. Despite the obvious results of steps taken, demands for the restoration of 'the environment' continues, but nobody knows or cares to determine the threshold level for a specific region's being considered environmentally degraded.

A future strategy could consider the following themes:

- Increasing water-use efficiency of agricultural and domestic user sectors, so that less water is drawn from the natural system and the least quantum of flow returns to the stream, and that only after treatment for reuse within the user sector, if possible.
- Scientifically assessing the loss of biodiversity due to water diversion for various competitive uses and restoring a minimum flow of water for good biodiversity plus some water for conservation.
- Assessing the effect of irrigated crop production and improved water use by humans and livestock on biodiversity, and analyzing the trade-offs.

There is also an acute need to estimate the benefit–cost ratio for water release to the environment (ecological economics), comparing not only 'before and after' conditions but also 'with and without' conditions in the scheme. We should stop comparing biodiversity loss in a seasonal or ephemeral stream with that of a perennial one, because ephemeral rivers do not constitute wetlands under Government of India rules. There is also a need to adopt Catchment Area Treatment (CAT), command area development, soil-moisture conservation, compensatory forests, social forestry and anti-pollution measures to minimize environmental flow needs.

Conclusions: addressing India's water management challenges

Like any ancient society, India has a self-balancing restraining system of constitution, water laws, policies, polity, media, institutions, organizations and standard practices, but with a silver lining of a society striving to rid itself of its negative baggage. But some parts of the developing world seem to be in the grip of perverse fundamentalism, applying a man versus nature approach to WRD. It will dissipate within a couple of decades, releasing the vast energies of the people in realizing their true potential to help the deprived sections of society in any of the water-use sectors: agriculture, human needs or nature.

Amendments have already been proposed to the existing laws, given their presently perceived inadequacies. Urgent actions are necessary on the adoption of a framework law, RBAs, water resource regulatory authorities and legal facilitation of inter-basin water transfer. Equity, sustainability and dependability provisions in WRD projects for each river basin – within and between basins – need a clear statement with reference to global interpretation. Principles of the judicious, optimal, least-cost combination of WRD from mega- to micro-scale for a river basin have yet to be woven into the effort. Climate change adaptation measures identified by the National Water Mission in the National Water Policy should be hastened.

The work of the Bureau of Indian Standards, with input from professionals, has to be accelerated. The principles and criteria for creation of RBAs, WRRAs and right down to water user associations need to be formulated. Model bills on water-related disasters (drought, flood, cyclones, desertification and landslides) ought to be enacted by the legislatures and implemented by governments.

Every facet of planning, design, administration and implementing agencies needs to be revisited. With climate change impending, India cannot afford failure. Priorities include setting up policy research centres and revising in-service training, merit-rating systems and transfer, promotion, award and reward policies. We should promote research and development, ethical values, quality control, quality assurance and periodic performance assessment of built infrastructure; ensure structural safety through field and laboratory assessment, followed by rehabilitation; create institutions for modernization or replacement under a code of practice; and improve coordination mechanisms to bridge these cadres and the polity for transformation.

The legislature, Election Commission and judiciary need to ensure that operations based on sound constitutional provisions are executed by a competent technocracy. Each political party ought to train its cadres to impart knowledge about government functioning to avoid costly mistakes. They should have special lessons on ethics in politics.

Governance and polity need to achieve synergy amongst all wings – social, economic, financial, legal, administrative, political, legislative, judiciary and media; enable the assessment of performance, and review policy periodically; and identify deficiencies in constitutional provisions, laws and standards. On their side, politicians ought to review party constitutions and internal democracy, remove feudalist tendencies, root out corruption with internal vigilance and watchdog bodies, root out favouritism, appeal to reason and rationalism, and weave together the interests of states for a common national programme.

Indian WRD professionals will have their hands full for another two or three decades to fast-track the many crucial issues mentioned in this paper. If 'business as usual' continues, problems are likely to multiply, hampering the progress already made in human development. Basic issues on the supply side urgently need to be resolved to include adaptation to likely climate change in the form of higher frequency of extreme events.

Disclosure statement

No potential conflict of interest was reported by the author.

References

Amerasinghe, P., Bhardwaj, R. M., Scott, C., Jella, K., & Marshall, F. (2013). *Urban wastewater and agricultural reuse challenges in India* (IWMI Research Report No. 147). Colombo, Sri Lanka: International Water Management Institute (IWMI).

Biswas, A. K. (1992). Indus water treaty: The negotiating process. *Water International, 17*, 201–209. doi:10.1080/02508069208686140

East India (Irrigation). (1903). Report of India irrigation commission 1901-03, Part II, Provincial, Presented to both Houses of Parliament by Command of His Majesty, East India (Irrigation), London.

Government of India. (1956). *Inter-state river disputes act 1956*. New Delhi: Ministry of Law, Justice and Company Affairs, Government of India.

Government of India. (1972). *Report of the irrigation commission 1972*. New Delhi: Ministry of Irrigation and Power. Retrieved July 12, 2017, from https://catalog.hathitrust.org/Record/009115425

Government of India. (1987). *National water policy*. New Delhi: Ministry of Water Resources, Government of India.

Government of India. (1999). *Report of the National Commission for Integrated Water Resources Development*. Volume I. New Delhi: Ministry of Water Resources, Government of India.

Government of India. (2002). *National water policy*. New Delhi: Ministry of Water Resources, Government of India.

Government of India. (2012). *National water policy*. New Delhi: Ministry of Water Resources, Government of India.

Government of India. (2015). *The constitution of India (As on 9th November, 2015)*. New Delhi: Ministry of Law and Justice (Legislative Department), Government of India.

Hardin, G. (1968). The tragedy of the commons, *Science*, New Series, *162* (Dec. 13), 1243–1248.

Kumar, M. D. (2014). The Hydro-institutional challenge of managing water economies of federal rivers: A case study of Narmada river basin, India. In D. Garrick, G. R. M. Anderson, D. Connel, & J. Pittock (Eds.), *Federal rivers: Managing water in multi layered political systems* (pp. 229–242). Cheltenham, UK and Northampton, MA, USA: Edward Elgar Publishing.

Ministry of Rural Development. (2007, October 31). *National rehabilitation and resettlement policy, 2007*. New Delhi: Department of Land Resources (Land Reforms Division), Ministry of Rural Development, Government of India.

Ministry of Water Resources, River Development and Ganga Rejuvenation. (2016, November 17) *Water sharing agreement with neighbouring countries*, Un-starred Question No. 260, Government of India.

Planning Commission. (2002, December 21). *10th Five Year Plan 2002-2007*. Volume I. New Delhi: Planning Commission, Government of India.

Sharma, Richa. (2017). Dam safety bill in cabinet soon, *The New Indian Express*, June 22. Retrieved from https://www.newindianexpress.com/nation/2017/jun/22/dam-safety-bill-in-cabinet-soon-1619407.html

The Gazette of India. (1956). *The river boards act, 1956, The Gazette of India*-Extraordinary, Part II, Section I, September 14.

India's water management debate: is the 'civil society' making it everlasting?

M. Dinesh Kumar and Chetan M. Pandit

ABSTRACT
This article discusses the bias of the growing constituency of civil society activists in India against conventional water management solutions implemented by the government, and the 'alternatives' they champion, which force the government to enter into an endless debate with these groups. The article goes into the fundamental reasons for this bias, and identifies four types of civil society activist: 'professional', 'ideologue', 'romantic' and 'doomsday prophet'. The article also argues that water bureaucracies in India should adopt evidence-based policy making, subjecting the 'alternatives' to the same degree of scrutiny as the conventional ones, to end the policy dilemma, while enhancing the overall quality of design, execution and management of projects for better outcomes.

Introduction

Many regions in India are experiencing water scarcity (Kumar, Sivamohan, & Narayanamoorthy, 2012). The demand for water from various competitive sectors of water use – domestic, industrial, livestock and irrigation – and for environmental management is growing as a result of growing population, higher growth in urban population, rising per capita incomes, changing consumption patterns and growth in manufacturing outputs (Amarasinghe, Shah, Turral, & Anand, 2007; Kumar, 2010). India needs to produce a larger quantity of different types of food grains, not only for direct consumption to feed the growing population but also as feed for livestock. Growing needs for fibre, sugar, milk, oil, pulses, etc. also need to be met (Alexandratos & Bruinsma, 2012). The rapidly expanding cities require more water, as per capita water demand for human uses is higher in cities as compared to rural areas. Rising per capita income changes lifestyle and consumption patterns, all impacting water demand, for producing food and for human uses (Amarasinghe et al., 2007; Kumar, 2010). Additional water will be required to keep the urban environments livable, for growing trees, maintaining gardens and recreational purposes. More importantly, nearly 60% of the population living in rural areas needs water for agriculture, a major source of their livelihood. So, there is no denying that developing water resources is the key to human development, economic growth and improving living conditions of the poor.

But there is hardly any consensus across stakeholders on what needs to be done to meet the growing challenges, in spite of the great urgency. Water is one theme which has strong social, economic and political ramifications, given its strong interlinkages with culture, society, politics and development. There are innumerable references to water as a social, economic, environmental and political good or even a strategic asset, as in Kautilya's *Arthashastra*, the ancient text on statecraft. The water management debate in India is much characterized by diametrically opposite views (Iyer, 2011; Kumar, 2010), so that in recent times the phrases 'pathology of bias' and 'pathological bias' have assumed immense significance in the discourse on water.

What makes the present scenario different from the past is the emergence of civil society (CS) advocates. The CS strongly campaigns for or against various policies related to water, energy and the environment. These policies have a huge impact on the nation's social fabric and the economy and therefore on its place in the international arena. Some even argue that the real objective of the CS advocates is to influence the economic agenda, and that water, energy and environment policies are only tools to achieve this goal. In any case, the discourse emanating from the CS groups influences the social and economic agenda, and that makes CS a political entity.

A water-sector professional has to take cognizance of the political overtones of CS advocacies. A close examination of the current debate on water resource development and management in India amongst scholars and practitioners clearly reveals a pathological bias of this class towards certain ideas, perhaps unique to the CS movement in our country. It often takes aggressive forms, though interestingly the perpetrators argue that they are opposed to any 'destructive' forms of development pursued by modern governments. Before we analyze its origin, we first list some of the ideas which have become targets of this pathological bias.

The pathological bias

Tap water vs. traditional systems for domestic use

It is reported that our ancestors in arid regions used to harvest rainwater and store it in tanks, often in the basement of houses, to be used for drinking and cooking. This is described at length in the books *Dying Wisdom: Rise, Fall and Potential of India's Traditional Water Harvesting Systems* (Agarwal & Narain, 1997), *Rajasthan Ki Rajat Boondein* (Mishra, 1996), and others. Some still idealize these systems as artefacts of ancient wisdom, with great relevance even in modern times, but now said to be 'dying'. What is ignored is the toll they took on people's health, and also the social fabric. The proponents of these systems are against the state's role in managing water for basic needs in rural areas, which, according to them, destroyed the traditional knowledge and wisdom that was instrumental in managing those old water systems by intervening in people's lives through building piped water supply schemes (based on Mishra ,1996, and several remarks of Rajendra Singh, the leader of Tarun Bharat Singh, which works in western Rajasthan, as reported by Gupta, 2011).

But the publications that extol rainwater harvesting, such as the publications of CPR Environmental Education Centre (http://www.cpreec.org/pubbook-traditional.htm), Agarwal and Narain (1997) and Mishra (1996), tell only half the story. What they do not discuss is that when traditional drinking water systems existed in Indian villages, water-borne infections wreaked havoc. Over 38 million people died from cholera, caused by the consumption of impure water, in India during the 100 years from 1817 to 1917 (Desonie, 2007). Cholera makes no distinction between the rich and the poor. So helpless were the people against this scourge that when the epidemic struck, the people used to flee the villages.

Cholera, gastroentiritis, infective hepatitis, etc. still exist, but are no longer dreaded epidemics that kill millions. This has been achieved only by chlorination and piped water supply (Chlorine Chemistry Council & Canadian Chlorination Coordination Committee, 2003). It is not that we have achieved all that is required; but the aim should be to serve all rural households with piped water supply schemes supplying treated water, and not revert to 'traditional systems', which provide untreated water.

Modern vs. traditional irrigation systems

It is said that our ancestors used to collect water for irrigation in ponds, tanks, *nadis* (traditional water-harnessing bodies of Rajasthan, very similar to tanks), and *talavs* (ponds) created by impounding small rivulets, and that these storages also served the purpose of recharging groundwater. The crops were irrigated from the water impounded by these storage structures, and from shallow open wells (Agarwal & Narain, 1997). These are now being touted as artefacts of our traditional wisdom, ignoring the social, political, cultural and technological context in which the people in those times depended on these systems, and more importantly, how far these irrigation systems were able to achieve food security.

It is common knowledge that some of these systems (tanks in South India, for instance) stood testimony to the increasingly extractive statecraft involving coerced labour, highly oppressive caste systems and the expropriation of surplus by elites, and were symbols of the enormous financial and physical power enjoyed by feudal landlords and warlords, respectively (Esha Shah, 2008).

Many tend to believe that Ancient India was a land of plenty. But the reality was quite different. India has always been a land of droughts, food shortages and famines. Some of the worst famines of the world were experienced in India. As many as 27 famines were recorded during the 110 years between 1770 and 1880, and more than 20 million lives were lost in some 20 famines that struck since 1850. Often one-tenth to one-third of the affected population was killed by starvation. Cannibalism was reported in some famines (Maharatna, 1996; Pandit, 2013).

The traditional irrigation systems failed not only during the devastating droughts but even in the best of times, and farming seldom led to prosperity. One drought was sufficient to drive a farming household into permanent debt. Practices such 'bonded labour' had their roots in this 'traditional system' of irrigation, which enabled nothing more than subsistence farming, except among the feudal landlords (Pandit, 2013).

With the advent of modern irrigation systems, which enabled large-scale transfer of water, first from the river to the farm, and then from naturally water-rich regions to water-scarce, drought-prone regions, plus intensive groundwater pumping from great depths in drought-prone regions, we are now able to achieve the twin objectives of producing sufficient food for a much larger population and generating a buffer stock to tide over food shortages during severe droughts; rural incomes have also increased (Pandit, 2013; Thatte, 2001). With mechanized transport, it became possible to shift large stocks of food grains to famine-affected areas. Though droughts occur even today, starvation due to famine has become a thing of the past. The challenge is to make all the drought-prone regions of India 'drought proof' and prepare them for the extreme climatic conditions which are characteristic of such regions.

Irrigated crops vs. rainfed crops

Some authors in the water sector believe that irrigated crops are not necessary to achieve food self-sufficiency in India (see e.g. Iyer, 2011). A group of development activists led by the late Shri R. R. Iyer, who proposed an alternative draft to the National Water Policy, argued that irrigation is the cause of most of the ecological problems. According to them, water in streams, rivers and aquifers should be left untouched (Iyer, 2011). The idea of environmental devastation caused by irrigation development is so sensationalized that *Pillar of Sand: Can the Irrigation Miracle Last?* (Postel, 1999) might be known to more people than any other book on irrigation or hydrology. They believe that the solution lies in rainfed crops. To them, these simple principles do not matter: "whether rain-fed or irrigated, crops have to take water from the hydrological system for the physiological process (i.e., transpiration), which comes from the soil through the roots" (van der Zaag, 2015, pp. 3–4); that to grow one kilogram of biomass of a given crop takes more water for transpiration under rainfed conditions than under irrigated conditions (Kumar, 2010); and that growing crops under rainfed conditions eventually depletes more water from the hydrological system and destroys more forests.

Up to almost the end of nineteenth century, agriculture was mostly rainfed, with little diversion of water from rivers and aquifers. The population to be fed was a fraction of what it is today, and yet India suffered perpetual food shortages for centuries (Thatte, 2001). However, the ardent supporters of rainfed agriculture do not seem to be bothered by this fact.

Canal water vs. groundwater

Some professionals have, over the years, developed a considerable bias against gravity irrigation systems. They believe that water which comes through open channels, supplied by the state with heavy subsidies, has little value and therefore should be put underground, later to be pumped out as 'groundwater', using energy also supplied by the state free of charge (Mukherji et al., 2009; Shah, 2009, 2011). They believe that canal systems worked in India only during British rule, because poor villagers could be coerced to grow exactly what their imperial masters in Britain wanted and pay the water charges on time (Shah, 2011). However, Ian Stone (1984), in his seminal work on canal irrigation in British India, had a different story to tell.

Stone noted that "the degree of control exercised by the canal irrigation authorities, and their manipulative powers with respect to agricultural improvements were effectively very limited" (p. 8). At the policy level, irrigation was linked to the multiple objectives of famine prevention, settling of unruly tribes, extended cultivation of cash crops along with expansion of cultivation, enhancing the taxpaying capacity of the peasants, and political stability. Hence, when it came to pricing canal water, it was difficult to define the objective clearly. The way irrigation water was delivered to the field was too dependent on indigenous bureaucratic and village-level structures. The prevailing concerns were of the peasant society to which the subordinate bureaucracy belonged, and not the Western concerns of efficiency and economy in water use. As a result, bribery, status, reciprocity, etc. asserted themselves (p. 9).

These professionals who attack public canal irrigation in independent India tooth and nail refuse to see the exemplary surface irrigation systems, in both the north and the south, built after India's independence, which have provided yeoman service to millions of farmers for several decades. According to them, the future of India's agriculture is well irrigation, which, in their view, is happening entirely thanks to 'private investments' (Mukherji, Rawat, & Shah, 2013; Nilekani & Subramanian, 2016; Shah, 2009). Ironically, they do not consider the mammoth sums the government spends for electricity subsidies to the farm sector, mostly for pumping groundwater, a worthy investment (Kumar, 2007). Hundreds of thousands of wells are failing due to groundwater depletion in the semi-arid, hard-rock areas (Anantha, 2009; Janakarajan and Moench, 2006; Kumar et al., 2012; Narayanamoorthy, 2015). However, while unleashing their attack on public investment in surface irrigation as one which yields no result in terms of increase in irrigated area, they do not want to look at this.

In their haste to write the obituary of publicly funded water resource schemes, they embark on the 'fifth labour of Hercules': cleaning up the mess in India's surface irrigation (see e.g. Shah, 2011). But the ideas they promulgate are no good, and on the contrary are capable of creating anarchy. One such idea is to divert water from streams and farmland to a private property called a 'well', a smart way of privatizing common property. They smartly cherry-pick examples, and indulge in preposterous and untidy statements, while using their poetic liberty to ridicule the government's decision to provide critical financial support to large projects that urgently need completion as the 'labour of Sisyphus', and public irrigation policy as the 'theatre of the absurd' (Shah, 2011, p. 11).

Large dams vs. check dams

Of all the pathological biases in the water sector, the bias against large reservoirs is the most severe. The CS in India is against construction of large dams and instead promotes construction of small check dams, thinking them ecologically benign and friends of farmers (D'Souza, 2002; McCully, 2003). They argue that large dams evaporate too much water, submerge too large an area and cause ecological devastation. They do not specifically say that, for the same quantity of water stored, check dams evaporate less water and submerge much less area – because to say so would oblige them to prove it, which they know they cannot do. Therefore, to sustain their argument, they avoid any comparison of the evaporation and the submergence per unit of storage of large dams versus small check dams.

However, the undisputable fact is that both evaporation and submergence per unit of storage are much larger for small dams than for large dams (Perry, 2001a). Therefore, to continue the tirade against large dams, a variety of detrimental effects of large dams are cited: large dams cause earthquakes; emit methane; destroy aquatic life; wipe out endangered species and habitats; trap sediment, which in turn causes morphological problems in deltas, and coastal erosion; and occasionally they collapse and drown people (McCully, 2003). Nowadays, religious sentiments are also used as a weapon in anti-dam protests, citing reasons like people worshiping the river and construction of a dam on the river offending their religious sentiments. For instance, one major objective of the Save Ganga Movement, which is against the building of proposed hydropower dams on the upper Ganges, is to "create mass awareness for an eco-friendly non-violent culture of development for the protection of our life-sustaining natural systems in general and of the sacred Ganga and the Himalayas in particular" (http://www.savegangamovement.org).

On the other hand, serious attempts to quantify the innumerable positive externalities that properly managed canal irrigation systems based on large dams offer society – better groundwater recharge, better quality of groundwater, lower energy cost of groundwater pumping, lower food prices, higher agricultural wages, etc. – are rare (Jagadeesan & Kumar, 2015; Shah & Kumar, 2008; Ranade & Kumar, 2004; Watt, 2008).

Creating permanent assets vs. digging holes

Some development scholars do not support permanent infrastructure in rural areas. They advocate *kachcha* roads (roads without an asphaltic or concrete surface), dug-out ponds, etc., so that employment can be manufactured after every good monsoon. The MNREGA (Mahatma Gandhi Employment Guarantee Act) scheme (Ambasta, Vijaya Shankar, & Shah, 2008), which has been spending roughly USD 5–6 billion every year for rural works since 2008–09 (Ministry of Rural Development, 2014), is an outcome of this. The Union government's budget outlay for NREGA schemes ranged from INR 300 billion in 2008–09 to a high of INR 401 billion (USD 6.3 billion) in 2010–11 and came down to INR 330 billion (USD 5.1 billion) in 2013–14 (Ministry of Rural Development, 2014). Apart from a strong desire to prove Keynesian economics correct, their immense liking for these earthworks seems to originate from their dislike for large infrastructure projects, which, in their view, always involve huge corruption at high levels. However, they ignore the fact that achieving transparency and accountability in such operations in rural areas is next to impossible, as one cannot even properly estimate the amount of earthwork involved, be it digging ponds or constructing roads (see e.g. Ambasta et al., 2008). It is an open secret that there is greater scope for corruption in 'hole digging' than in building highways, bridges and dams. But they seem to have greater faith in local self-governments, such as the Panchayati Raj. Hence, they follow the difficult path of doing 'social audits' on the performance of these structures (see e.g. Shah, 2012). It would be interesting to see how many of these ponds and embankments continue to exist, even one or two years after construction (assuming they were actually constructed).

The report of the Committee for Revision of MNREGA Guidelines (Shah, 2012) is eloquent about management information systems, social audits, voluntary disclosure by *panchayats*, involvement of CS, inclusion of new activities such as watersheds in the list of works, financing NREGA, etc., and only wants to address governance issues. But there are serious conceptual flaws in the very design of the scheme (Kamath, 2010; Kumar, Bassi, Sivamohan & Vedantam, 2011; World Bank, 2011) that are creating diseconomies from the point of view of water management (Bassi & Kumar, 2010; Kumar et al. 2011; World Bank, 2011).

Pricing of water vs. free water

Some researchers in the water sector have a distaste for ideas such as water pricing (De Fraiture & Perry, 2002; Perry, 2001b), water tax, property rights in water, etc. (Shah, 2009, 2010). Some believe that it has never worked anywhere in achieving water demand management, and that there are serious theoretical and practical issues with metering and pricing (De Fraiture & Perry, 2002; Perry, 2001b) or imposing tax for water, and that the worldwide experience with these instruments is just a work in progress (Shah, 2009, 2010). The counterparts of these researchers in CS and political dispensation think that water should be free, so that the poor can obtain it at no cost. They refuse to recognize that while all the water which comes from precipitation is a free good, the dams which store it, the motors which lift it, and the pipes which convey it are not. The activists and the politicians know very well that in such situations, it is one's social status, and financial and political clout, that determine access to water, not one's willingness to pay for it. These activists and politicians even assume that those families which use less than 700 litres of water per day would eventually all be classed as poor, irrespective of the family size, a point also made by other scholars recently, who argue that such policies only benefit the rich (see e.g. Narayanamoorthy, 2014). They even ignore the fact that the ultra-poor families living in cities do not have water connections at their dwellings, and therefore would get no benefit from water's being 'free'.

In contrast to this, in the developed countries, utilities resort to public awareness and education to regulate water use and prevent wastage only after using instruments such as water pricing, water tax and pollution tax to influence consumer behaviour (Hassel, 2007; Syme, Nancarrow, & Seligman, 2000). Unfortunately, in India we begin and end with awareness and education programmes and shy away from trying harder, politically inconvenient options.

'Dirty power' vs. 'clean energy'.

The current fad is using only 'clean energy' to solve the problems of climate change and global warming. There is nothing wrong with this as a goal, except that now only solar power is accepted as clean. Everything else is dirty. Thermal pollutes the air with carbon, and emits greenhouse gases; nuclear has the potential to kill thousands of people, plus there is the problem of disposal of radioactive waste; hydro dams submerge precious forests. Wind energy, once a favourite of the environmental NGOs, is no longer clean because it kills birds; windmills are erected on forest land; the routes to windmill locations are through forests; etc. Many of the claims are themselves disputable, and the numbers which are touted are certainly disputable. But, here also, they shy away from looking at the hard facts.

The hard fact is that one kilowatt of solar power requires 10 m^2 of land; one megawatt requires 1.0 ha of land; and therefore 1000 MW requires 1000 ha of land. The enormous cost of this required land will be a non-issue only if we can put solar farms on saline tracts or deserts which are non-reclaimable. There is a lot of resistance if anyone comes out with empirical arguments on the energy economics to beat their claims. One such argument is that solar power costs many times more than hydropower and even thermal power, if we look at the cost per kilowatt hour of energy (Bassi, 2015).

More importantly, there are serious questions about how clean this 'clean energy' is. Solar photovoltaic panels generate electricity at a very low voltage, and only intermittently, depending on momentary local insolation. Therefore, the energy has to be first stored in battery packs. The battery uses sulphuric acid and lead and has to be replaced every three years. But, being well-mannered and climate-conscious citizens, we are not supposed to ask such uncomfortable questions, typical of a bourgeois mindset.

There are still more fallacious ideas. One of them is to replace India's five-million-plus diesel pumps (used for irrigation) with solar pumps to find a permanent solution for the poorly managed groundwater–energy nexus. According to them, this is economically very viable for poor farmers with an 85% subsidy. A five-horsepower solar pump costs nearly INR

500,000. The farmer needs to shell out just INR 75,000 (USD 1200), the other INR 425,000 (USD 7000) coming from the government (Shah, Verma, & Durga 2014; Kishore, Shah, & Tewari, 2014). It is clear that the full economic cost of the system nowhere figures in the benefit–cost analysis; instead, only the private costs and benefits for the farmers are considered. From the sheer scale of investment required, one does not need a sharp mind to perceive who would be the beneficiaries of this technology. A simple calculation shows that to replace only 100,000 pumps would cost the exchequer INR 42.5 billion (USD 660 million); five million wells would mean INR 2.12 trillion (USD 33 billion).

But the real problem is not in managing such large sums of money from the state exchequer but the fact the plan is not economically viable, even if we consider the positive externality of reduction in carbon emissions, while completely ignoring the land costs (Bassi, 2015). Furthermore, the past experience of solar photovoltaic panel–based energy systems in rural areas is very poor. From plain vandalism to theft – the charge controller is yanked off, the battery is stolen – the systems rarely last more than a couple of years. And there are serious questions about the maintenance requirements and the availability of technical skills in the rural areas (Nathan, 2014).

But, these problems notwithstanding, there are even ideas of farmers producing excess power (from equipment that was subsidized in the first place) and selling it to the grid to make money (Shah et al. 2014). It is one thing to subsidize a system to help a farmer produce electricity for his or her own use in the farm; but it is totally unwise to build infrastructure like feeder lines to the power grid entirely at public expense to bring down the cost of generation and supply of electricity for private individuals who could then make a profit selling power to the energy market.

The pathology of bias

Now that we have discussed pathological bias, let us look at the pathology of this bias. A clinical examination of the background of these dogmatists would point to not one but several causes, some of which are systemic in nature.

The easiest to decipher is the existence of four types of CS activists: 'professional', 'ideologue', 'romantic' and 'doomsday prophet'. The professional sees environmental activism as source of livelihood and more. Just as one can earn a livelihood by building irrigation systems, one can earn a livelihood and more by opposing irrigation projects.The ideologue believes that large dams and modern irrigated agriculture are bad. The very word 'ideology' excludes any explanation of *why* these are bad.

The romantic dreams of an India where all people live in beautiful little villages, in their beautiful little huts, tilling their (half-hectare) fields, and engaged in occupations such as beekeeping or bamboo-mat-making, while their little children play with little goats and sheep. Questions like 'What is the annual worldwide demand for honey, or bamboo mats?' are *persona non grata*.

The doomsday prophet thinks that the modern way of living will eventually push this planet into a catastrophe. Such thoughts are reinforced by the 'precautionary principle', which, restated in simple English, means that the absence of incontrovertible scientific proof that large dams will do no harm irrefutably proves that large dams will do immense harm. The doomsday prophet lives in perpetual fear that any action will trigger some catastrophe, and ends up embracing 'do-nothing-ism'.

To this straightforward typology must be added some more complex types.

First, many are those who have insufficient knowledge (in for example hydrology, engineering, economics, sociology, development or political science) to even comment on any of these complex issues, and their ability to make objective assessments of what is good and what is bad can be open to challenge. But, interestingly, these self-proclaimed water experts question all the scientific knowledge accumulated from several centuries of research and application in the related disciplines, using numbers and arguments which are only reflective of their personal bias and prejudice rather than reality (see e.g. Iyer, 2011; Singh, 2001). The underlying reason is a tendency towards attention seeking, no matter that implementing these fancy ideas would impair development. When such questioning comes from individuals who work with the establishment, people tend to listen to them with seriousness. Unfortunately, the institutional regimes responsible for scrutinizing the knowledge generated in the society for scientific validity do not appear to be strong and are only weakening over time.

Otherwise, which society would believe that a series of small check dams will provide more water at less cost and with less submergence, and thus can be an alternative to large dams? When the activists who were opposing the Sardar Sarovar Project were asked, "What is the alternative?" they had no answer. However, after the publication of *Dying Wisdom* (Agarwal & Narain, 1997), suddenly, they had an answer: the ponds and tanks, and everything our ancestors did. As explained at length in the foregoing section, this answer overlooked the fact that these systems did not serve our ancestors' needs.

Second, there are strong vested interests in maintaining the status quo with regard to technologies, practices and implementation approaches in the water sector. For these vested interests, replacing village ponds with piped water supply, building large reservoirs and enabling intensive irrigated farming instead of rainfed subsistence agriculture, and building power plants for rural lighting means little work for their establishments, as their rhetoric will be of hardly any use in the new scenario. Perhaps this tendency also has to do with the culture of criticism and playing it safe, rather than innovating with new technologies and approaches and running the risk of failure.

Third, there are some organizations which knowingly or unknowingly play into the hands of some foreign entities, who are their benefactors, and whose primary interest is to see that India remains on the edge, does not make sustainable investments in the water and energy sectors, and instead keeps depending on them for basic things like food. The insurmountable criticism from the environmental lobby in the late 1980s and 1990s against Green Revolution technologies and large irrigation systems, which were pivotal in making India self-sufficient in food, completely forgetting the 'ship to mouth' past, is just one case in point (Shiva, 1991). Shiva (1991) calls the adoption of Green Revolution technologies and the building of large dams for irrigation in India an act of violent aggression against ecology, culture and society.

Fourth, there are many institutions, mostly small and grass-roots in nature, which really do great community service work using water as the vehicle of development, but which often fall prey to the politics of the big players in the sector, who operate globally, nationally and locally.

Lastly, as pointed out by Pandit (2014), some CS activists view democracy as a system where the 'communities' are the ultimate owners of natural resources, and all the power of decision making regarding their development and use should be vested in them, or in village-level local self-government. Therefore, they perceive the state as an aggressor that takes control of natural resources from local people and interferes with their freedom by imposing decisions on them. These perceptions morph into debates of 'centralized versus decentralized', the State vs. the People, etc. But, obviously, state agencies are required for building large thermal power plants and big dams; local village *panchayats* cannot handle such massive works. To get over the problem of accepting this reality, the CS argues against such projects. They even believe that a *gram sabha* (a body of citizens in a village, who have voting rights, recognized by Indian Constitution as a legitimate body as per the Panchayati Raj Act of 1993) is capable of deciding the environmental flow of a river, a complex concept that has challenged hydrologists and environmental scientists for over two decades.

Unfortunately, often the functioning of the government also reinforces the above beliefs. The rehabilitation and resettlement in large water resource projects is often tardy; compensation is not disbursed in time; the work is of poor quality; projects are taken up and not finished; many projects underperform; and there is corruption at every level. The CS organizations use such instances to create stereotype images. Such attempts succeed because the common man does not understand the flaw in deriving inferences from small samples. When one Tawa project (in Madhya Pradesh) causes waterlogging, all irrigation projects are painted as 'environmentally disastrous'. If some farmer suicides are reported in the Bhakra command in Punjab, the unfortunate incident is gleefully used to blame large dam–backed canal irrigation.

Conclusions

Water management challenges for countries like India are greater than ever before. Not only are water demands growing, but there is growing competition for the available water. Increasing climate variability and extremes will add to the challenge. A multipronged approach will be required to manage water in future, comprising demand management, including management of intersectoral allocation (Kumar, 2010; Saleth & Dinar, 1999); transfer of water from surplus regions to regions of scarcity to enhance not only the utilization potential of the available water but also the use value of the resource (Kumar, Singh, & Sharma, 2005; Kumar, Malla, & Tripathy, 2008; Kumar, 2010); and increasing the multiannual storage of water to meet the challenges induced by climate variability. Water infrastructure, together with institutions, policies and regulations, are the basic requirements to meet these challenges, especially in developing countries, where much of the growth in water demand is taking place (Tortajada & Biswas, 2015).

It would be presumptuous to say that science and technology can solve all of today's water problems. All problems do not even necessarily have a solution. But there is no doubt that the best solution can emerge only from science and technology. The country can find these solutions, provided we do not politicize the problem too much. To begin with, we need to create a corpus of innovative ideas and solutions, which should be based on the science and technology as they exist at a given point of time, which take into account not only the needs but also the aspirations of 1.3 billion people and which are supported by well-trained hydrologists, engineers, economists, sociologists and environmental scientists. 'Barefoot water engineers' and 'parahydrologists' have a role and should have a respectable position in society, but they cannot replace qualified technocrats, just as paramedics cannot replace surgeons. Opposition to large projects is not a sufficient qualification to become an environmentalist, and certainly not a water expert. Nor can it replace a formal degree in environmental sciences. But certain influential groups believe that they can cut the Gordian knot of water and energy crisis by vehemently arguing that the big systems are inefficient and only local-community-based solutions will work.

As someone has put it, we cannot predict our future, but we can create one. We have already created a gloomy future for water. It is certain that India will face acute water scarcity, and many areas in central, peninsular and western India are already experiencing it, in different manifestations. There is no *if* about water scarcity in times to come. The question is, other than observing a World Water Day once a year, what are we doing about it? When future historians write about India's water problems, very likely one of the verdicts will be: India experienced acute water scarcity not because the technocrat community implemented some wrong decisions but because the CS engaged them in endless discussions and court battles, kept the 'discussion pot' boiling eternally, and stopped the technocrat community from implementing any worthwhile decisions.

Water management involves complex considerations of physical, technical, social, economic, legal, institutional and cultural aspects, and hence policy making in that field is also going to be a difficult process. The appropriateness of these policies depends largely on the evidence on which they are based. Here, both quantity and quality matter. The quantity of evidence concerns the number of regions and localities from which it is collected. The quality of evidence concerns the scientific rigor with which it is analyzed. Many times in the recent past, the decision to pursue a particular paradigm in water management was not supported by sufficient scientific evidence of its effectiveness but by a simplistic understanding of the underlying hydrological and geohydrological processes (Perry, 2013). Hence, there is a dilemma in policy circles. The government's failure to use the mounting evidence of the effectiveness of the paradigms and strategies it followed in the water sector over the past 6–7 decades to achieve food security, water security and rural development, and subject the non-constructive alternatives proposed by CS to rigorous scrutiny, has also added to this dilemma.

Often, there is inadequate application of sound concepts and principles of water management in the water resource projects of India (Perry, 2013). At times, there is also a problem in translating good water management concepts into robust project designs, executing projects in a cost-effective and time-bound manner and maximizing positive outcomes and minimizing negative ones, because of the failure on the part of the government to change the archaic rules and procedures, and revamp outdated institutions.

Disclosure statement

No potential conflict of interest was reported by the authors.

References

Agarwal, A., & Narain, S. (1997). *Dying wisdom: Rise, fall and potential of India's traditional water harvesting systems*. New Delhi: Centre for Science and Environment.

Alexandratos, N., & Bruinsma, J. (2012). *World agriculture towards 2030/2050: The 2012 revision* Global Perspective Studies Team, ESA Working Paper 12-03. Rome: Agricultural Development Economics Division, FAO.

Amarasinghe, U. A., Shah, T., Turral, H., & Anand, B. K. (2007). *India's water future to 2025–2050: Business-as-usual scenario and deviations* Research Report 123. Colombo, Sri Lanka: International Water Management Institute.

Ambasta, P., Vijaya Shankar, P. S., Shah, M. (2008). Two years of NREGA: The road ahead, *Economic and Political Weekly*, Insight, *XLIII* (8), 41–50.

Anantha, K. H. (2009) Downward dividends of groundwater irrigation in hard rock areas of southern peninsular India Working Paper # 225. The Institute for Social and Economic Change, Bangalore.

Bassi, N. (2015) Irrigation and energy nexus: Solar pumps are not viable. *Economic and Political Weekly*, March, 2015.

Bassi, N., & Kumar, M. D. (2010). NREGA and rural water management in India: Improving the welfare effects Occasional Paper # 3. Institute for Resource Analysis and Policy, Hyderabad.

Chlorine Chemistry Council & Canadian Chlorination Coordination Committee. (2003). *Drinking water chlorination: A review of disinfection practices and issues*. Retrieved June 10, 2016, from http://www.waterandhealth.org/drinkingwater/wp.html, downloaded on.

D'Souza, D. (2002). *Narmada dammed: An enquiry into the politics of development*. New Delhi: Penguin Books India.

De Fraiture C., & Perry C. (2002) Why is irrigation water demand inelastic at low price ranges? Paper presented at the conference on irrigation water policies: micro and macro considerations, 15-17 June 2002, Agadir, Morocco.

Desonie, D. (2007). *Hydrosphere: freshwater systems and pollution*. New York, NY: Chelsea House.

Esha Shah, E. (2008). Telling otherwise: A historical anthropology of tank irrigation technology in south India. *Technology and Culture, 49*, 652–674.

Gupta, S. (2011). Demystifying 'tradition': The politics of rainwater harvesting in rural Rajasthan. *India, Water Alternatives, 4*, 347–364.

Hassel, T. (2007). *Promoting behaviour change in household water consumption: A literature review, prepared for smart water*. Victoria, Australia: URS Australia Pvt. Ltd.

Iyer, R. R. (2011) National water policy: An alternative draft for consideration. *Economic and Political Weekly, XLVI*, 26–27. June 25, 2011.

Jagadeesan, S., & Kumar, M. D. (2015). *The sardar sarovar project: Assessing economic and social impacts*. New Delhi: Sage Publications.

Janakarajan, S. and Moench, M. (2006). Are wells a potential threat to farmers' well-being? Case of deteriorating groundwater irrigation in Tamil Nadu. *Economic and Political Weekly, 16* (September): 3977–87.

Kamath, R. (2010). National Rural Employment Guarantee Act: An effective safety net? *IIMB Management Review, 22* (1): 42–55.

Kishore, A., Shah, T., & Tewari, N. P. (2014) Solar irrigation pumps: Farmers' experience and state policy in Rajasthan, *Economic and Political Weekly, XLIX* (10), 55–62.

Kumar, M. D. (2007). *Groundwater management in India: Physical, institutional and policy alternatives*. New Delhi: Sage Publications.

Kumar, M. D. (2010). *Managing water in river basins: Hydrology, economics, and institutions*. New Delhi: Oxford University Press.

Kumar, R., Singh, R. D., & Sharma, K. D. (2005). Water resources of India. *Current Science, 89*, 794–811.

Kumar, M. D., Malla, A. K., & Tripathy, S. K. (2008). Economic value of water in agriculture: Comparative analysis of a water-scarce and a water-rich region in India. *Water International, 33*, 214–230.

Kumar, M. D., Bassi, N., Sivamohan, M. V. K., & Vedantam, N. (2011). Employment guarantee and its environmental impacts: Are the claims valid? *Economic and Political Weekly, XLVI* (34), 69–71.

Kumar, M. D., Sivamohan, M. V. K., & Narayanamoorthy, A. (2012). The food security challenge of the food-land-water nexus in India. *Food Security, 4*, 539–556.

Maharatna, A. (1996). *Book for famines: The demography of famines: An Indian historical perspective*. New Delhi: Oxford University Press.

McCully, P. (2003) Big dams, big trouble, *New Internationalist. People, Ideas and Action for Global Justice, 354*, March 1.

Ministry of Rural Development. (2014). Mahatma Gandhi rural employment guarantee act 2005 report to the People. New Delhi: Dept. of Rural Development, Ministry of Rural Development, Government of India, 2nd February, 2014.

Mishra, A. (1996). *Rajasthan ki Rajat Boondein* [Rajasthan's silver drops]. New Delhi: Environment Division, Gandhi Peace Foundation.

Mukherji, A., Facon T., Burke J., de Fraiture C., Faurès J.-.M., Füleki B., ... Shah T. (2009). *Revitalizing Asia's irrigation: To sustainably meet tomorrow's food needs*. Colombo, Sri Lanka: International Water Management Institute; Rome, Italy: Food and Agriculture Organization of the United Nations.

Mukherji, A., Rawat, S., & Shah, T. (2013). Major insights from India's minor irrigation censuses: 1986-87 to 2006-07. *Economic and Political Weekly, XLVIII* (26 & 27), 115–124.

Narayanamoorthy, A. (2014) Is free water supply bad economics? *Business Line*, January 2, 2014.

Narayanamoorthy, A. (2015). Groundwater depletion and water extraction cost: Some evidence from South India. *International Journal of Water Resources Development, 31*, 604–617.

Nathan, H. S. K. (2014). Solar energy for rural electricity in India: A misplaced emphasis. *Economic & Political Weekly, 49*, 60–67.

Nilekani, R., & Subramanian M. (2016) Revitalise our aquifers: India can be water secure, but for that we need to act urgently. *Times of India*, May 30, 2016.

Pandit, C. (2013). Overview of water resources development in India, lecture note prepared for the Training Programme. on "Water Insecurity and Climate Variability: Theoretical Understanding and practical Programming," February 26-March 02, 2013.

Pandit, C. (2014). Environmental overenthusiasm. *International Journal of Water Resources Development*, 30 (1), 110–120.

Perry, Chris.J. (2001a). World commission on dams: Implications for food and irrigation. *Irrigation and Drainage, 50*, 101–107.

Perry, Chris.J. (2001b). Water at any price? *Irrigation and Drainage, 50*(1), 1–7.

Perry, Chris.J. (2013). Beneath the water resources crisis, discussion. *Economic and Political Weekly, XLVIII*, 59–60.

Postel, S. (1999) *Pillar of sand: Can the irrigation miracle last?* A World Watch book. Washington DC: W W Norton & Company Inc.

Ranade, R., & Kumar, M. D. (2004) Narmada water for groundwater recharge in north Gujarat: Conjunctive management in large irrigation projects, *Economic and Political Weekly*, 39 (31), 3510–2513.

Saleth, R. M., & Dinar A. (1999) *Water challenge and institutional responses (a cross country perspective)* Policy Research Working Paper Series 2045, World Bank, Washington D.C.

Shah, T. (2009). *Taming the anarchy: Groundwater governance in south Asia*. Washington D.C: RFF Press.

Shah, T. (2010) *India's ground water irrigation economy: The challenge of balancing livelihoods and environment*. New Delhi: Central Ground Water Board, downloaded on. Retrieved March 28, 2015, from www.cgwb.gov.in/documents/papers/incidpapers/Paper%203-%20Tushaar%20Shah.pdf

Shah, T. (2011). Past, present and the future of canal irrigation in India. *India Infrastructure Report 2011*. Mumbai: Infrastructure Development Finance Corporation.

Shah, M. (2012). *The report of the committee on revision of operational guidelines of MNREGA* final report submitted to the Ministry of Rural Development. New Delhi: Government of India.

Shah, Z., & Kumar, M. D. (2008). In the midst of the large dam controversy: Objectives. *Criteria for Assessing Large Water Storages in the Developing World, Water Resources Management, 22*, 1799–2814.

Shah, T., Verma, S. and Durga, N. (2014). Karnataka's smart, new solar pump policy for irrigation. *Economic and Political Weekly, XLIX* (48): 10–14.

Shiva, V. (1991). *The violence of green revolution: Third world agriculture, ecology and politics*. London and New Jersey: Zed Books Ltd.

Singh, R. (2001). Johad: A structure of community self-reliance. The 27th Bhaikaka Memorial Lecture, 16th Indian Engineering Congress, IIT Kharagpur.

Stone, I. (1984). *Canal irrigation in british India perspectives on technology change in a peasant economy*. Cambridge: Cambridge University Press.

Syme, G., Nancarrow, B. E., & Seligman, C. (2000). The evaluation of information campaigns to promote voluntary household water conservation. *Evaluation Review, 24*, 539–578.

Thatte, C. D. (2001) Water resources sector: AN overview and prospects for the 21st Century. 14thDr AN Khosla Memorial Lecture, 16th Indian Engineering Congress, IIT Kharagpur, December 01-04, 2001.

Tortajada, C., & Biswas, A. K. (Eds.). (2015). *Water infrastructure*. London, United Kingdom: Routledge, Taylor and Francis.

Watt, J. (2008) The effect of irrigation on surface-ground water interactions: Quantifying time dependent spatial dynamics in irrigation systems Thesis Submitted to Charles Sturt University for the Degree of Doctor of Philosophy, School of Environmental Sciences, Faculty of Sciences, Charles Sturt University, March 2008.

World Bank. (2011). *Social protection for a changing India, volume II*. Washington, DC: World Bank.

van der Zaag, P. (2015). *Soil and water management for rainfed agriculture in semi-arid areas, securing livelihoods and food production by slowing the water flow in catchment areas.* Delft: UNESCO-IHE Institute for Water Education, June.

Proposing a solution to India's water crisis: 'paradigm shift' or pushing outdated concepts?

M. Dinesh Kumar

ABSTRACT
This article is a critique of the report of the committee chaired by Dr Mihir Shah on restructuring the Central Water Commission and Central Ground Water Board of India. It shows that the recommendations of the committee are not based on any sound understanding of the federal nature of water administration in India, water-sector performance or the problems confronting it. The 'paradigm shift' in the suggested approach to water management is based on flawed analysis of the performance of surface irrigation systems and outdated concepts of irrigation efficiency, and reflects the professional bias of its members against large water infrastructure and wishful thinking about what schemes like aquifer mapping can achieve.

Reforming India's water sector

India's water sector is facing insurmountable problems. In this country with 1300 million persons to feed, and with rapid urbanization and industrialization and changing consumption patterns and life styles, the challenges facing water managers are immense. Managing water today is no longer only about developing new sources through conventional means by constructing reservoirs, digging wells and laying canals and pipelines, but also about finding new sources of water and allocating the limited water amongst various competitive uses, while protecting the hydrological integrity of our catchments, rivers, lakes and aquifers. Inter-sectoral water allocation requires greater use of sound economic principles for efficient pricing, introduction of water-use restrictions, etc. Water resource management requires application of ecological sciences, ecological economics and environmental economics. It is quite obvious that our water-sector institutions have to be equipped with more technical manpower, with greater competence and with people from multiple disciplines. They also call for new institutions for basin-wide water allocation and for undertaking resource management action.

It also needs to be appreciated that because water is a state subject in India (except for the inter-state river basins), most of the reforms have to happen at the state level, in order to affect changes in the orientation and working of the agencies which plan, design, execute and run water projects. The role of the central-level agencies such as the Central Water Commission (CWC), Central Pollution Control Board and Central Ground Water Board (CGWB) is limited to hydrological monitoring of rivers (for river discharge and sedimentation); flood forecasting; groundwater survey and assessment (recharge estimation and groundwater quality monitoring); and water quality monitoring of aquatic systems. In other words, their role is advisory in nature, and they have no direct stake in the outcomes of their decisions. For instance, the CWC and CGWB can work jointly to come up with a basin management plan for an inter-state river basin, for the larger interest of the basin water economy. But it is up to the affected parties (in this case the riparian states) to agree to this plan. It is quite possible that one state has to forgo some of its economic interests for the benefit of another. As a result, in the current institutional set-up, the state may show no interest in such plans, as they are not statutory in nature.

Mihir Shah committee recommendations

A committee set up by the Ministry of Water Resources in December 2015, under the chairmanship of Dr Mihir Shah, a member of the former Planning Commission of India, was to come out with specific recommendations for restructuring the CWC and CGWB. This was on the premise that these institutions are quite outdated and their work needs to be made more relevant to the changing context of India's water sector. The committee seems to have ignored certain basic facts about the working of India's water administration, particularly the centre–state relationships, in suggesting the restructuring of these two agencies. The committee, whose report, 'A 21st Century Institutional Architecture for India's Water Reforms', is now available on the ministry's website (http://wrmin.nic.in/writereaddata/Report_on_Restructuring_CWC_CGWB.pdf), made several recommendations based on its own diagnosis of the ailing sector.

Some of the problems identified by the committee are: (1) the efficiency of public irrigation schemes is as low as 35%; (2) there is a mounting gap between potential created and potential utilized in the irrigation sector, to the tune of 26 million ha; (3) groundwater is a "golden goose", which could cater to the water needs of India's future generations and which accounts for two-thirds of India's irrigation, but its use is alarmingly unsustainable; (4) the proportion of area irrigated by canals is declining fast; and (5) there is no scope for further development of surface water in the country, particularly since most rivers in Peninsular India are highly developed and already facing severe environmental water stress.

The committee's recommendations, however, are mere reflections of the strong ideological bias and position of its members collectively rather than outcomes of any proper diagnosis of India's water-sector problems. This is evident from the fact that the committee has left out the vast body of scientific literature on India's water problems, the functioning and performance of the water-sector agencies and what could be done to mitigate the problems. Some of the recommendation of the Committee are: (1) securing 26 million ha of additional irrigation by closing the gap between irrigation potential created and potential utilized, which it considers easily achievable with minor investments in completed and ongoing public irrigation projects; (2) water demand management through promotion of water users' associations (WUAs) in irrigation commands; (3) participatory groundwater management through nationwide aquifer mapping, to be completed urgently; (4) river rejuvenation; and (5) integrated water resources management, practised by taking the river basin as the unit for planning and management.

A moratorium on large dams

The committee's suggestions for putting too many caveats on future dam building by calling it the greatest source of human tragedy in the country in the form of displacement, and over-reliance on groundwater as a future source of water for all needs, are not based on any realistic assessment of the situation on the ground. The problem with groundwater is that in the regions which have plenty of groundwater (like in the eastern Gangetic Plains), the demand for water for agriculture and other sectors is low, owing to very small landholdings, high rainfall and humidity, and several agro-ecological problems, such as flooding; on the other hand, in the areas where the demand for water for agriculture and other sectors (urban drinking, industry) is excessive and unmet, the groundwater resources are over-exploited, and millions of wells are failing every year (Kumar, Sivamohan, & Narayanamoorthy, 2012). Clearly, the committee has not understood the implications of this dichotomy for the feasible options for meeting future water-scarcity problems.

The committee goes on to suggest that CWC and CGWB work under a single umbrella body, named the National Water Commission, with their present chairpersons as members, along with a few other members and wings for ecology, environment, etc., and headed by a person with public administration background – a bureaucrat. The committee assumes that such an institution can promote integrated water resources management at the basin level, for both surface water and groundwater, and take their water management agenda forward. What the committee failed to understand is that the reason for lack of integrated planning is not only the lack of coordination and data sharing, but lack of ability to foresee how future development of the resource is going to take place in different sectors and lack of control over that development, which still lies in the hands of the concerned state departments.

As is clear from the several interviews of its chairman, Dr Mihir Shah, the committee thinks that this would mark a major paradigm shift in India's water sector, and would have long-lasting outcomes and impacts. Contrary to this claim, the fact, as also suggested by international experience, is that most of the solutions being proposed by the committee for India's water sector can bring about only cosmetic changes. Even if one accepts for a moment that these recommendations, bundled under the 'new paradigm of water management', are capable of bringing about major reforms in the sector, the larger concern remains: Can the two central agencies (CWC and CGWB) be held responsible for the poor state of affairs in the water sector and also held accountable for implementing the new water management paradigm, as envisioned in the report, for the committee to justify the restructuring proposal? The report does not provide any analytical base to establish the link between the two. In the process, the committee has taken the wrong patient to the operating table. Ideally, institutional reforms are required to effect changes in the functioning of the state water agencies which plan and develop water resources. But the committee has not made suggestions for improving their working, and instead wants to restructure the central agencies.

False premise, faulty diagnosis

In short, the entire report is built on a false premise, a faulty diagnosis and misrepresentation of the facts. First, the paradigm shift being advocated is based on neither a proper analysis of the problems nor an understanding of what needs to be done to rescue the ailing water sector in India. Instead, it echoes some of the outdated concepts used for judging performance of irrigation systems and some of the stale ideas repeatedly tried for almost two and a half decades by lending agencies in India and many other developing countries with no significant positive outcomes. Second, the two agencies whose work is under review (CWC and CGWB) are not linked to the problems identified by the committee, nor are they parties in implementing the consequent recommendations for solving them, as they are made out to be. We will deal with them one by one, and discuss them in the subsequent paragraphs.

The committee uses time series data on (net) area irrigated by canals and tanks as a percentage of total (net) irrigated area, instead of the absolute area irrigated by surface sources, to comment on the performance of (public) surface irrigation, and concludes that surface-irrigated area has declined over time in India (Shah, 2016, p. 25). This analysis is flawed and misleading, as the denominator, i.e. the "total (net) irrigated area in the country", dramatically increased over the time period considered (1950–64 to 2005–09). Further, the committee conveniently ignored the fact that the area reported under "other sources" is mainly the area irrigated through lifting from public canals and that it should be included in the surface-irrigated area. Also, the committee did not take cognizance of the fact that many large cities and towns in India receive water from public reservoirs, which are primarily meant for irrigation.

Thus, using concocted data on "percentage area under surface irrigation", the committee tries to build the argument that the era of dam building in India is over, as according to it the returns from continued public investments in irrigation projects have been negative during the past few decades, primarily due to incredibly low irrigation efficiencies (25–35%) and low utilization of the newly created potential. The committee cites this as the major reason for reorienting the CWC, which in the committee's opinion is more equipped to plan and design water resource development projects for supply management, given that due to the "current focus of the CWC on 'development' of water resources for supply-side management alone, the CWC is staffed exclusively with engineers drawn from the Central Water Engineering Group 'A' Service, and lacks almost totally a capacity in any other discipline that interfaces with water resource management" (p. 78).

How inefficient are the public irrigation systems?

First of all, the efficiency of public irrigation is far higher than what is reported, if one assesses it at the basin scale (Perry, 2007, 2013). International agencies have long been using the concept of basin water-use efficiency (see e.g. Keller & Keller, 1995). A mere 25–35% efficiency of public irrigation schemes, as noted by the committee, means that nearly 70% of the water released from the reservoirs or diversion systems would be lost. This works out to around 280 km^3 of water annually. This 'wastage' should end up in the natural sink that is the rivers and the oceans. If we acknowledge the fact that most of the large irrigation systems are in the water-scarce states of India (Gujarat, Maharashtra, Andhra Pradesh, Tamil Nadu, Karnataka and Telangana), most of this water should appear at the last drainage point of the rivers in these states. However, this does not happen. Most of these rivers have no significant amount of water draining out in normal years, as noted even by the committee. Hence, the efficiency argument is completely flawed.

Is India's existing irrigation potential under-utilized?

As regards the committee's remarks on the large gap between 'irrigation potential created' and 'potential utilized', these terms are found only in the administrative lexicons of India and not used in irrigation science. Moreover, the way estimates of 'irrigation potential created' are arrived at is nothing but fallacious. These figures are often unrealistic, heavily inflated, as they are based on the estimates of quantum of water available in the reservoir and a 'design cropping pattern' which never happens in reality. Water inflows into reservoirs can change depending on the rainfall in the catchment and many upstream developments. Also, farmers shift to water-intensive crops once irrigation water is made available, further lowering the 'irrigation potential utilized' (Kumar, 2016). A modern-day water scientist with a good understanding of hydrology and socio-economic aspects would never resort to such numbers to critique an agency's performance (Perry, 2013).

On the other hand, as regards the potential utilized, there is heavy under-reporting of the actual area irrigated by canals, with no accounting for water lifted from canals and drains by engine owners, and the area irrigated by wells in the command which benefit from the seepage of canals and return flows from gravity irrigation (Kumar, 2016). This also means that we need to review the way data on canal-irrigated area are collected. In sum, it is not possible to achieve an additional 26.0 million ha of irrigation, and the gap between the irrigation potential created and potential utilized is just a misnomer. Every drop of water in these water-scarce basins is captured and used within the basin, though some scope exists for reducing non-beneficial uses of water, such as evaporation from barren soil or fallow.

What can WUAs achieve?

The report strongly argues for promoting and strengthening WUAs in canal command areas to address the problems of poor water-use efficiency and the wide gap between potential created and potential utilized. But WUAs will have too little a role in achieving water demand management, except taking care of the distribution issues to some extent at the level of the tertiary canals. As a matter of fact, WUAs are defunct in all the states, in the absence of devolution of any kind of powers to them. The state irrigation departments are not willing to share any of their powers.

Water demand management requires efficient pricing of water in irrigation and other sectors, and rationing or fixing volumetric entitlements in water. These are long overdue. Only such measures can bring about improvements in water-use efficiency, through optimal use of irrigation water for the crops, allocation of the available water to more efficient crops, or saving water in the existing uses and selling it to alternative uses at a high price. It has long been understood that reallocation of water to alternative uses can effect demand reduction in the inefficient use sectors, particularly agriculture, by creating incentive for water-use efficiency improvements, and water-use efficiency improvements alone cannot achieve demand reduction in the agriculture sector in most situations (Kumar & van Dam, 2013). But these are political decisions, which need to be executed by the respective state agencies, and are in no way within the purview of the agencies whose work is under scrutiny. The committee has not delved into these challenging issues but simply found refuge in the 'success' of implementing irrigation management transfer programmes in different parts of the world documented by international agencies such as FAO (Food and Agriculture

Organization of the United Nations) and the erstwhile International Irrigation Management Institute (see e.g. Garces-Restrepo, Vermillion, & Muñoz, 2007; Vermillion, 1997).

Is canal-irrigated area in India declining?

As noted by Darrell Huff in *How to Lie with Statistics* (1991), "The secret language of statistics, so appealing in a fact-minded culture, is employed to sensationalize, inflate, confuse, and oversimplify." This is precisely what was done by the committee when it compared surface irrigation with well irrigation. It used figures of "percentage of net area irrigated by different sources" to show that canal irrigation has declined and well irrigation risen consistently. But if one looks at the actual gross irrigated area by different sources, it clearly shows that the gross canal-irrigated area, after steadily going up, has stagnated in recent years, and the same has happened to well irrigation, though well-irrigated area is much greater. There are many reasons for this phenomenon of stagnation of area under irrigation, including those discussed in the previous section. But one major point is that there is growing reallocation of water from large reservoirs to cities for water supply as cities expand (Mukherjee, Shah, & Kumar, 2010).

Leaving aside the issue of misuse of statistics, such crude statistical comparisons should have been avoided by a committee which strongly considers groundwater and surface water part of the same hydrological system and advocates integrated surface and groundwater management. Showing its clear bias against large dams, the committee makes no mention of the fact that large reservoirs (most of which were primarily built for irrigation, excluding those for hydropower) today supply water to several large cities, a factor which would have been considered by any scholar who makes an objective assessment of the sector. Not doing that shows the professional bias against public irrigation.

Is aquifer mapping a panacea for groundwater over-exploitation?

On the groundwater side, it is a well-established fact that the two major reasons for over-exploitation of aquifers are the absence of well-defined water rights in groundwater and inefficient pricing of electricity supplied to the farm sector. The problem is surely not due to lack of sufficient information about occurrence of groundwater and its flows. The farmers as well as official agencies know well that the resource is fast depleting in many pockets. What is lacking is effective institutions which can allocate the available groundwater amongst users and regulate the use of the resource through proper monitoring to achieve sustainability. Participatory aquifer mapping can do little to halt this ongoing menace. As envisaged, it can provide micro-level details of the groundwater-bearing formations (good aquifers) within the geological strata in different localities (Kulkarni & Vijay Shankar, 2009; Shah, 2013). The committee is unable to visualize how participatory aquifer mapping gets translated into participatory groundwater management under the much-touted National Aquifer Management Program, with a budgetary allocation of INR 35.39 billion (USD 532 million) under the 12th Five-Year Plan.

While we can invest more resources in refining the current assessment methodology, we already have sufficient information to initiate management actions in the problem areas. But those actions are going to be institutional in nature, and what we lack is the political will from the state governments to do it. This is quite clear from the lack of enforcement of groundwater acts despite a number of states having these on paper, such as the Maharashtra Groundwater (Development and Management) Act of 2009 (Government of Maharashtra, 2013), which was enacted on 3 December 2013. The experience of developed countries such as the United States and Australia in dealing with groundwater management issues clearly shows that the solution lies in creating robust institutions which can clearly define the water rights of individual users and enforce them or institute a tax on groundwater use based on withdrawal volume.

Ignoring the role of water infrastructure?

The problem today is that most of the basins in the naturally water-scarce regions are 'closed', with no water going uncaptured. But agriculture growth is suffering in these regions due to lack of water. Telangana, Rayalaseema and western Rajasthan are just a few examples (Amarasinghe et al., 2004). Cities in these basins are not getting enough water for their rapidly growing populations, as they face severe competition from the agriculture sector, which is already under stress. The rivers in these regions need water to maintain ecosystem health. While the committee discusses "rejuvenation of rivers" at length, it fails to offer any practical suggestions on how to achieve it, except talking platitudes about integrated surface and groundwater development. But to get water back in the river and to maintain the base flows from aquifers, we need to cut down water withdrawals drastically. Given the precarious water balance in many river basins, one way to achieve these multiple objectives is through inter-basin water transfers from water-abundant basins, which is already happening in limited ways in different basins that are characterized by sharp differences in resource endowments.

We would require much larger and more sophisticated water infrastructure for inter-basin water transfers – to store, divert and transfer water from water-rich basins to distant regions that are perennially water-starved. Therefore, to take up the new challenges, capacity must be built in state and central-level institutions in disciplines such as hydrology, hydraulics, groundwater modelling, water engineering, river morphology, environmental hydrology, dam safety, ecological and environmental economics and water law. There is no doubt that water demand management must receive great attention in the coming years, with the cost of production of water reaching astronomical heights. But to achieve this, we need legitimate regulatory institutions which can look at issues of water allocation and water pricing at the state level, with good understanding of the economics of water.

Capacity building of CWC and CGWB officials in the field of integrated water resources management is welcome. There is also scope for broadening the thematic areas of the training to cover environmental, economic and social aspects. But the real question is how such training modules from the National Water Academy, the training wing of CWC, would find takers in the state water agencies, which do not show any willingness to introduce water pricing or to experiment with new approaches to basin planning and institutional models for water resources management.

Further, asking the National Water Academy to build the capacities of farmers and NGOs to undertake water management activities in the field is expecting too much. The report says:

> It is not enough to train water professionals (engineers) within the CWC and in the states. Water resource management has to be a participatory process with all stakeholders (including NGOs)

and local communities (particularly farmers at the *panchayat* level), who should be made aware of the water resource issues and who need capacity building for management of the limited water resources (e.g., participatory irrigation management, appropriate crop selection, micro-irrigation, conjunctive use, wastewater reuse/recycling, etc.). The present training activities of the NWA in this respect are practically negligible. (p. 85)

This is the mandate of the water and land management institutes set up in different states nearly three decades ago. These institutions undertake farmer training on land management, crop management and farm water management for strengthening command area development (Government of India, 2013).

Conclusions

Integrated water resources management is also a complex model of water governance and management. The challenge is to operationalize the concept through appropriate coordinating institutions at the river basin level. As discussed by Biswas (2008, p. 12), "Operationally it has not been possible to identify a water management process at a macro- or meso-scale which can be planned and implemented in such a way that it becomes inherently integrated, however this may be defined, right from its initial planning stage and then to implementation and operational phases." This is in the realm of the state water agencies which have to create such coordination mechanisms. These challenges are time-dependent. The committee did not deal with these myriad issues. Instead, it passes judgements that these organizations have outlived their mandate, but recommends restructuring these agencies into a National Water Commission with basin-wise headquarters, without any vision of future water management needs and the role they have to play to make it a reality. This can only demoralize the cadre of scientists and engineers in these institutions.

The Harvard historian David Blackbourn, in his recent book, *The Conquest of Nature: Water, Landscape and the Making of Modern Germany* (2007), documents how each succeeding generation takes the achievements of the past generation for granted and wonders how their predecessors could have been so unwise as to not have dealt with the new generation of challenges! The state of the art (of water management) is always provisional. Finally, contrary to the committee's claim, which is reflected in the title of its report, there is no 'twenty-first-century architecture' in the institutional model suggested in its report, but only wishful thinking. To do the former, the committee should have incorporated eminent experts on institutional economics and organizational behaviour in its subcommittees. Instead, it co-opted many civil society activists, most of whom are known for their strong ideological position against large water resource systems, in these subcommittees.

Disclosure statement

No potential conflict of interest was reported by the author.

References

Amarasinghe, U. A., Sharma, B. R., Aloysius, N., Scott, C., Smakhtin, V., & de Fraiture, C. (2004). *Spatial variation in water supply and demand across river basins of India* Research Report 83. Colombo, Sri Lanka: International Water Management Institute.

Biswas, A. K. (2008). Integrated water resources management: Is it working?. *International Journal of Water Resources Development, 24*, 5–22. doi:10.1080/07900620701871718.

Blackbourn, D. (2007). *The conquest of nature: Water, landscape and the making of modern Germany*. New York and London: W. W. Norton & Company.

Garces-Restrepo, C., Vermillion, D., & Muñoz, G. (2007). *Irrigation management transfer programme: Worldwide efforts and results* FAO Water Report 32. Rome: Food and Agriculture Organization of the United Nations.

Government of India. (2013, December). *Guidelines on command area development and water management programme*. New Delhi: Ministry of Water Resources, Government of India.

Government of Maharashtra. (2013). *Maharashtra Groundwater (Development and Management) Act*, Maharashtra Act. No. XXVI of 2013, Maharashtra Government Gazette, 3rd December 2013.

Huff, D. (1991). *How to lie with statistics*. New York: Penguin.

Keller, A. A., & Keller, J. (1995). *Effective efficiency: A water use efficiency concept for allocating freshwater resources*. Arlington VA: Centre for Economic Policy Studies, Winrock International.

Kulkarni, H., & Vijay Shankar, P. S. (2009). Groundwater: Towards an aquifer management framework. *Economic and Political Weekly*, Commentary, February 7, 2009, 13–17.

Kumar, M. D. (2016). Irrigation sector 'turn around' in Madhya Pradesh? *Economic and Political Weekly, 51*, 67–70.

Kumar, M. D., & van Dam, J. (2013). Drivers of change in agricultural water productivity and its improvement at basin scale in developing economies. *Water International, 38*, 312–325. doi:10.1080/02508060.2013.793572.

Kumar, M. D., Sivamohan, M. V. K., & Narayanamoorthy, A. (2012). The food security challenge of the food-land-water nexus in India. *Food Security, 4*, 539–556. doi:10.1007/s12571-012-0204-1.

Mukherjee, S., Shah, Z., & Kumar, M. D. (2010). Sustaining urban water supplies in India: Increasing role of large reservoirs. *Water Resources Management, 24*, 2035–2055. doi:10.1007/s11269-009-9537-8.

Perry, C. J. (2007). Efficient irrigation, inefficient communication, flawed recommendations. *Irrigation and Drainage, 59*, 367–378. doi:10.1002/ird.323.

Perry, C. J. (2013). Beneath the water resources crisis. *Economic and Political Weekly, 48*, 59–60.

Shah, M. (2013). Water: Towards a paradigm shift in the twelfth plan. *Economic and Political Weekly*, Special article, January 19, 2013, *XLVIII*, 40–52.

Shah, M. (2016). *A 21st century institutional architecture for water reforms in India*. Final Report submitted to the Ministry of Water Resources. New Delhi: River Development & Ganga Rejuvenation, Government of India.

Vermillion, D. L. (1997). *Impacts of irrigation management transfer: A review of the evidence*. Research report 11. Colombo, Sri Lanka: International Irrigation Management Institute.

Water transfer from irrigation tanks for urban use: can payment for ecosystem services produce efficient outcomes?

L. Venkatachalam and Kulbhushan Balooni

ABSTRACT
Many Indian states have begun to transfer water meant for irrigation to non-agricultural purposes, but the economic and environmental consequences are not adequately understood. Transfer of water out of water bodies from rural areas not only reduces the economic welfare of the traditional water users but also reduces their incentives to manage these water bodies on a sustainable basis. The study explores the possibility of introducing the mechanism of 'payment for ecosystem services' at the grass-roots level in the Indian context as a return for reallocation of water from irrigation to urban uses so that it can produce a non-zero-sum outcome for villagers, farmers, urban consumers and governments.

Introduction

Acquisition of agricultural land for development projects in India has prompted political unrest in many states and produced economic consequences detrimental to sustainable agricultural development at micro level. Such acquisition reduces economic welfare, especially for the unwilling farmers (Chaudhry, 2015; Ghosh, 2012; Levien, 2012; Ramachandraiah & Srinivasan, 2011) and landless agricultural labourers and sharecroppers (Narain, 2009; Vij & Narain, 2016), if they are driven away from their principal source of livelihood. The welfare loss is high if the affected parties lack alternative economic opportunities. The literature on land acquisition does not focus on this welfare loss, which either destroys or degrades water resources and their ecosystem services. In addition to land acquisition, governments and private entities transfer water out of rural water sources for urban uses (Chandran, 2016; Gandy, 2008; Narain, 2009), mostly without compensating the affected. What institutional mechanism is required that can address the consequences of this scenario?

This study analyzes the consequences of transfer of water for urban use from water bodies traditionally used for irrigation purposes and then explores *payment for ecosystem services* (PES) to address the economic trade-offs associated with water transfer. This is achieved by analyzing the case of irrigation tanks in the state of Tamil Nadu in India, which are historically well known for their contribution to agriculture development in the state.

The article first discusses the process of water transfer from traditional irrigation sources for non-agricultural uses and its possible economic consequences. This is followed by an overview of challenges in conservation and management of irrigation tanks in Tamil Nadu, including an appraisal of large-scale transfer of water from irrigation tanks for non-agricultural purposes and how the current institutional challenges in tank management can be overcome by introducing PES. It then describes the concept of PES for water usage and the ramifications of its implementation based on empirical literature. Finally, this study analyzes conditions for implementing water PES in the Indian context, followed by conclusions.

Water transfer for non-agricultural uses and its economic consequences

Traditionally, water for urban use in India has been drawn largely from rivers, lakes, large reservoirs and subsurface sources. Due to urbanization, state governments are transferring water from traditional irrigation sources such as large reservoirs and tanks in rural areas, for the benefit of urban consumers. Such interventions by the state can hurt water users in rural areas. While the issues associated with land acquisition are being widely debated and discussed in both political and academic spheres (Ghatak, Mitra, Mookherjee, & Nath, 2013; Singh, 2012), the economic and environmental impacts of water transfer do not get adequate attention in the existing literature on water management in the Indian context. Water is acquired in two different ways. First, when irrigated land is converted to development purposes, under either forced or voluntary arrangements, the water rights attached to the land ownership are also taken away from the farmers. As a consequence, a variety of water-based non-market ecosystem services enjoyed by different stakeholders in rural areas are also lost. Second, when water is directly transferred from irrigation sources for high-value urban use, the water users in rural areas are deprived of various non-market ecosystem services associated with irrigation water. In both situations, the economic value of welfare loss is not accounted for anywhere in the system.

Water sources generate multiple ecosystem services (Barbier et al., 2008; TEEB, 2010) that are used by farm and non-farm users across wide geographical regions. These services are of four types: *provisioning services,* which include food, freshwater, fibre, fuel, and biochemical and genetic materials; *regulating services,* consisting of climate regulation, hydrological regulation, water purification and wastewater treatment, erosion regulation and regulation of pollination; *cultural services,* comprising spiritual, inspirational, recreational, cultural and educational values; and *supporting services,* in the form of soil formation and nutrient recycling (Millennium Ecosystem Assessment, 2005).

Ecosystem services contribute both directly and indirectly to the economic welfare of innumerable stakeholders: farmers benefit directly in economic terms from irrigation water use; rural households use water for various extractive and in-stream uses; landless households rearing ducks and livestock are (indirectly) dependent on irrigated agriculture, especially during the post-harvest period. Environmental economists (e.g. Kolstad, 2011) have identified *non-use values,* such as *option value, quasi-option value* and *existence value,* which generate utility for individuals and households, expressed in terms of either their willingness to pay to preserve these benefits or willingness to accept compensation for forgoing them. The non-use values become positive whenever households attach cultural and religious values to some unique water source, for instance, temple tanks in rural areas of India. The nature and the size of all these benefits differ across different water users and different water sources, and also across climatic conditions (Kumar, Panda, Niranjan, & Bassi, 2012).

Water-based ecosystem services have certain unique characteristics: (a) most of them have high economic value but do not have a market price; (b) some are tangible and some are intangible; (c) some are visible and some are invisible; (d) some are objective while others are subjective; and (e) some are local while others are global. Therefore, identifying and quantifying the economic damage due to loss of ecosystem services from water reallocation is hard. Since the compensation to water users is linked to the underlying property rights, the users are rarely compensated; they have no well-defined ownership rights over water. De jure, the government is the custodian of water resources. Yet it can, in many cases, transfer irrigation water without bothering about the effects of such an act on the current users. In such situations, the resulting welfare loss is significant if the lost ecosystem service does not have any substitute and if the users of that ecosystem service belong to an economically vulnerable group. Water can become a non-renewable resource locally for short time frames due to reasons such as monsoon failure and climatic variations. Therefore, diverting water from a local source without due consideration for its future availability will impose social costs not only on the present generation but also on future generations.

The urban water utilities divert water from distant irrigation tanks and lakes to cities as water bodies around these cities have either dried up or disappeared due to encroachment for other activities (Seenivasan & Kanagavalli, 2014). In many situations, water from existing sources has become insufficient for the growing needs of ever-increasing urban populations (Kumar, 2014). Though the supply of freshwater to cities from their traditional sources is declining for a variety of reasons, the demand for water is increasing, leading to an ever-widening supply–demand gap. Hence, cities have to resort to diversion of water from far-away sources primarily meant for agricultural uses. In Tamil Nadu, diversion of irrigation water for urban use is becoming common. For example, over 180 million litres of water is being drawn every day from the Veeranam tank, a large irrigation tank with a catchment area of 25 km^2 in the state, to meet supplementary water requirements in the Chennai Metropolitan Area; the tank is 145 km south of Chennai (http://chennaimetrowater.gov.in). In addition, many water supply augmentation schemes currently being implemented in the Chennai Metropolitan Area have resulted in water withdrawal in bulk from neighbouring irrigation sources such as the Palar River, approximately 82 km south of Chennai, with a basin area of 18,300 km^2 and potentially supplying 4200 m^3 of water per year for agricultural, industrial and household uses in the basin area. This water transfer deprives the farmers of water from these sources. With the increasing urban population, water diversion from the neighbouring irrigation sources is likely to accelerate in the coming years, potentially at the cost of economic welfare for diverse water users.

In a water-scarce economy, diverting water from a low-value irrigation use to high-value urban uses results in efficiency gains and can produce beneficial effects in the aggregate. In a strict economic sense, the maximum benefits from water diversion can be achieved only if the marginal benefits of an extra unit of water allocated across all users are maximised. With careful and systematic planning, a significant quantity of water from the agriculture sector can be diverted to non-agricultural purposes without significant loss in agricultural production, productivity or profit (Cullet, Bhullar, & Koonan, 2015; Molle & Berkoff, 2006). For example, taking some water out of over-irrigated or waterlogged areas can reduce soil salinity and raise crop yields, increasing the net marginal benefits in the concerned sectors.

Table 1. Number of tanks and tank irrigated area in Tamil Nadu, India.

	Tanks			Area irrigated by tanks (thousand ha)	
Year	With command area of 40 ha or more	With command area of less than 40 ha	Total number of tanks	Gross area irrigated	Net area irrigated
2001–02	7,529	31,837	39,366	607	537
2002–03	7,529	31,837	39,366	461	422
2003–04	7,529	31,837	39,366	419	385
2004–05	7,933	32,386	40,319	504	465
2005–06	7,933	32,386	40,319	641	575
2006–07	7,982	33,278	41,260	569	531
2007–08	7,982	33,278	41,260	546	506
2008–09	7,984	33,278	41,262	580	540
2009–10	7,984	33,278	41,262	534	503
2010–11	N/A	N/A	N/A	573	533
2011–12	N/A	N/A	N/A	567	528

Source: Season and Crop Report of Tamil Nadu, 2001–02 to 2011–12, Department of Economics and Statistics, Government of Tamil Nadu, Chennai.

Over the years, the gross irrigated area and net irrigated area under tank irrigation in Tamil Nadu have declined significantly (Narayanamoorthy, 2004; also see Table 1) for reasons such as farmers giving up agricultural activities, or agricultural land being used for non-agricultural purposes. As water previously used for cultivation is now available for other uses, efficiency can be gained if the government adopts a systematic approach to identify and divert such unutilized water for urban use.

Since drinking water gets top priority in India's National Water Policy (MoWR, 2012), governments find the traditional irrigation sources easy targets for diverting water for urban use. Nonetheless, an approach like the one currently being adopted by government agencies to divert irrigation water can certainly cause distortions in water supply among rural households that are already experiencing water scarcity in many parts of Tamil Nadu, which also results in economic losses from agricultural and allied activities such as duck rearing and animal husbandry. Moreover, many irrigation tanks are not being properly maintained, so water diversion from such tanks will be devastating for agriculture and allied activities. Since water scarcity is a key factor of distress among many Indian farmers at present (Reddy & Mishra, 2009), the command-and-control method of water transfer, with governments solely deciding how much water is to be transferred and from where without consulting the stakeholders, would make the condition of the farmers much worse.

Challenges in conservation and management of irrigation tanks

Tanks have been an important source of irrigation since ancient times in India, but empirical evidence suggests that over time the land area irrigated by the tanks has declined significantly (Palanisami, Meinzen-Dick, & Giordano, 2010). According to Narayanamoorthy and Suresh (2017), the area irrigated by tanks across India declined from 4.56 million ha in 1960–61 to 2.04 million ha in 2010–11, and the share of tank irrigation in the net irrigated area fell from 18.49% to about 3% during this period. Narayanamoorthy and Suresh also point out that encroachment of tanks for agricultural and urban use, collapse of traditional village institutions that managed irrigation tanks efficiently, and a shift in the preferences of the farmers from tank irrigation to tube-well irrigation are the reasons for the declining trend in tank irrigation.

In Tamil Nadu, the cultivable area irrigated by tanks fell from 38% of the gross cropped area in 1960–61 to 19.47% in 2000 (Narayanamoorthy, 2004). On the other hand, reports by the Department of Economics and Statistics of the Government of Tamil Nadu for the period 2001–02 to 2011–12 show that while the total number of tanks increased, the area irrigated by them declined (Table 1). This is because the farm ponds created by the farmers in their agricultural lands were also included in the estimation of the number of tanks by government officials (A. Narayanamoorthy, personal communication). However, anecdotal evidence suggests that not only has the area under tank irrigation shrunk but also the number of traditional irrigation tanks has declined, especially at the regional level. For example, of a sample of 1350 tanks in Thiruvallur and Kancheepuram Districts, 90 were abandoned and 210 were completely encroached upon (Lakshmi, 2013). Most of these water bodies served as drinking water sources as well.

The reasons identified by Narayanamoorthy (2007) for the decline in the number of tanks include the spread of tube-well technology, change in land-use patterns in the catchment area, heavy siltation of tanks due to negative externalities of upstream activities (e.g. over-grazing), encroachment of catchment areas, changing patterns of rainfall, poor governance, fractured village institutions, changes in land ownership, and conversion of agricultural land under the tank command to non-agricultural purposes. Specifically, tanks and lakes closer to cities experience irreversibility problems arising from discharge of sewage and industrial effluents, dumping of urban solid waste and encroachment by industrial and commercial establishments (Venkatachalam & Jayanthi, 2016).

To cope with increasing water scarcity, the Tamil Nadu government has taken steps to restore and rehabilitate the traditional irrigation tanks. For instance, 569 tanks governed by the state's public works department and 80 ex-*zamindari* tanks – tanks managed by rich landlords from the end of the eighteenth century till 1956 – were rehabilitated and modernized with assistance from the European Union, with a financial outlay of INR 1793.90 million (USD 1 = INR 64.07 as of 15 May 2017). These tanks serve a command area of 73,161 ha (Sakthivadivel, Gomathinayagam, & Shah, 2004). Similarly, the government rehabilitated various system and non-system tanks through the World Bank–assisted Irrigated Agriculture Modernization and Water Bodies Restoration and Management Project at the cost of INR 25,470 million (http://tniamwarmtnau.org/). Despite these measures, a major question that remains unanswered is: how can these rehabilitated tanks be managed on a sustainable basis?

A larger part of the recent literature on tank management claims that failure of the traditional institutions governing irrigation tanks led to their poor performance at present (Palanisami et al., 2010). How to fix institutions that can not only revive the tanks but also manage them on an intertemporal basis has become a serious policy question. On the other hand, the existing literature that emphasizes the importance of issues related to engineering and sociological aspects of tank management seems to downplay other aspects, such as hydrology of the tank systems, e.g. how runoff into the tank is affected by land-use patterns in the catchment and command areas, how increased groundwater exploitation in the upper catchment areas reduces inflows into the tank, etc. (see e.g. Kumar, Vedantam, Bassi, Puri, & Sivamohan, 2012; Lele, Patil, Badiger, Menon, & Kumar, 2011). This implies that institutional interventions to conserve and manage irrigation tanks need to address the above issues concerning tank hydrology rather than merely looking at the civil engineering works of development and management of tanks and their command area.

The institutions governing irrigation tanks determine efficiency, equity and sustainability in the use of tank water. However, identifying appropriate institutions and making them work in the field is a challenging task. A study by Sakthivadivel et al. (2004) showed that each of the best performing tanks has its own techno-institutional mechanisms (such as norms for desilting or repairing feeder canals and tanks) for water management, and the success of the traditional institutions depended largely on a decision-making process that addressed the concerns of all stakeholders; the final decisions reached through consensus were accepted by all as 'fair'. This study found that lack of cohesiveness, non-inclusiveness and faction-ridden institutions cause low performance of the tanks. Shah and Raju (2001) found in the context of Rajasthan tanks that lack of commonality of interests among key stakeholders – the command area farmers, tank-bed farmers, fishermen and other user groups – led to poor performance of rehabilitated tanks. Jegadeesan and Koiji (2011) attributed the deterioration of village tanks to disappearing caste hierarchy in rural areas, which once played a crucial role in efficiently managing those tanks. That is, the people belonging to lower castes in rural areas were responsible for managing irrigation and agriculture while the upper caste people owned the majority of the cultivable land; when the educational and economic status of the lower-caste people improved over a period of time, these people moved from agriculture to non-agricultural activities, and therefore tank management gradually deteriorated.

From the review of the existing literature, it emerges that the irrigation tanks are already experiencing management-related problems in the wake of increased demand for the limited tank resources, and hence water transfer from the tanks for urban use may not be feasible unless or until new institutional arrangements are put in place to strengthen collective action and improve the performance of the tanks.

Studies on water governance pertaining to irrigation tanks focus mainly on whether 'participatory irrigation management' can help improve their performance (Bhatt, 2013; Pant, 2008). For example, Pant (2008) argued that the modern water users' associations (WUAs) that are established by the governments through political process of empowerment of users can fill the vacuum created by the end of traditional village institutions, which successfully managed the irrigation tanks in the past. He prescribes creating such associations as an enabling environment for tank management. Similar prescriptions are reported in other studies as well (e.g. Mukherji et al., 2010). Since such studies selectively choose only a handful of successful cases, Meinzen-Dick (2007) found it doubtful whether the institutional panaceas manufactured from such results can be replicated and proven effective elsewhere. Moreover, a closer look at these institutional prescriptions reveals that they are based on a lack of understanding of how institutions evolve in the socio-ecological domain.

The evolutionary theory of institutions predicts that efficient institutions replace the inefficient ones irrespective of the nature of their outcomes (North, 1990). What is actually happening in the tank command area is what this theory predicts. A significant number of farmers wanted to minimize the transaction costs of surface irrigation, which was unreliable, and they moved from tank irrigation to bore-well irrigation. This is because the marginal benefits (as against the costs) of bore-well irrigation are relatively higher (whereas those of tank irrigation are lower), due to its reliability, and the transaction costs of bore-well irrigation are relatively lower (whereas those of tank irrigation are higher), although the initial investment in bore-well irrigation is high. Therefore, when the number of farmers available for collectively maintaining the tanks falls below a tipping point, the tank management system collapses. In this context, Kumar, Vedantam, et al. (2012) devised certain protocols for choosing tanks for rehabilitation so that the benefits of using the rehabilitated tanks are significantly higher than the costs of creating and sustaining the tank management institutions. The catchment area of the tank must yield sufficient water for inflow, and only tanks with catchments having low well density and a high 'command area to wetland area ratio' should be used for rehabilitation. They further suggested the use of a rainfall-runoff model to assess tank inflow (in case the catchment area underwent major land-use changes), and realistic estimation of water demand for irrigation in the command area (see also Kumar & Rao, 2017).

The discussion above suggests that governments merely creating and deploying WUAs in an institutional vacuum, without taking into account the economic and institutional dynamisms in both agriculture and rural areas, would result in social costs exceeding social benefits. Unlike traditional institutions, the roles of the WUAs created by the governments are not well defined. These WUAs suffer from the problems of elite capture and non-cooperation (Reddy & Reddy, 2005). Moreover, when an irrigation tank can provide large-scale, multiple ecosystem benefits, WUAs with a narrow irrigation management objective could trigger conflicts among various stakeholders using other ecosystem benefits. So, restoring collective action requires bringing all major stakeholders under one umbrella and deploying appropriate incentive-based institutional mechanisms that can favourably change the relative benefits and costs of overall tank management. In addition to government, communities and WUAs, a market-based institutional mechanism, namely PES, can play a significant role in implementing an effective governance system to manage tanks for multi-purpose use in general, and for water diversion for urban use in particular.

Payment for ecosystem services in the water sector

PES has been successfully implemented to manage certain critical environmental resources in general and water resources in particular around the world (Clements et al., 2010; Engel, Pagiola, & Wunder, 2008; Lipper, Sakuyama, Stringer, & Zilberman, 2009; Locatelli, Rojas, & Salinas, 2008; Pagiola, 2002; Wunder, 2008). In north-eastern France, for instance, Nestlé Waters, which sells Vittel, a mineral water brand, has been successfully compensating farmers for adopting best practices in dairy farming (measures such as abandoning agrochemicals, composting animal waste and reducing animal stocks) to improve the quality of raw water obtained from the catchment areas of the Vosges Mountains (Perrot-Maitre, 2006). In Bolivia, an in-kind compensation programme encourages upstream farmers to protect cloud forests and provide water services to an international donor agency interested in conservation and to the downstream farmers, who benefit from dry-season water flows (Asquith & Wunder, 2008). In the Central American region, Costa Rica implemented PES (*pago por servicios ambientales* in Spanish) to compensate land owners for implementing sustainable forest management plans so that an increase in the hydrological services such as groundwater recharge, along with benefits that include reduced greenhouse gases and increased biodiversity, could be accomplished (Pagiola, 2002, 2008). Mexico's payment for hydrological services (e.g. groundwater recharge), implemented in different segments of the forest areas, aims at conserving the forests to maintain the quantity and quality of water (Fisher, Kulindwa, Mwanyoka,

Turner, & Burgess, 2010; Munoz-Pina, Guevara, Torres, & Brana, 2008). In some parts of Mexico, PES increased the participation of especially the poor in conservation activities, helping reduce poverty (Alix-Garcia, de Janvry, Sadoulet, & Torres, 2008).

In South Africa, the Working for Water programme, a version of PES, has been fruitfully implemented to restore mountain catchments to enhance streamflow to downstream water bodies. Clearing of invasive alien species in the catchments and riparian zones, which impaired streamflow, was taken up by certain municipalities, and the clearing cost was met through government subsidies, and payment by water utilities and water consumers (Turpie, Marais, & Blignaut, 2008). Water charges were imposed on water consumers for control of invasive alien species, planning and implementation of a catchment management programme, pollution control, demand management and water use control. Working for Water has been hailed as a most successful integrated land and water management programme, since paying for improved environmental services brought palpable benefits to different stakeholders (Turpie et al., 2008). In Uganda, PES played a critical role in protecting wetlands and enhancing their ecosystem services to support the livelihoods of a significant number of poor households (Nalukenge, Antle, & Stoorvogel, 2008). In China, two nationwide programmes – the Sloping Land Conversion Programme and the Forest Ecological Services Compensation Fund – have incorporated payment for water services to protect major river basins against siltation and floods (Huang, Upadhyaya, Jindal, & Kerr, 2009). Programmes that use payment for water services are being implemented in various developing and industrialized countries (Schomers & Matzdorf, 2013).

India is also planning to introduce market-based instruments in a significant way in select areas of environmental management (MoEF, 2006). There is a scope for adopting PES to protect and allocate some of India's critical environmental resources, including water (Behera, Mishra, & Nayak, 2011). Indeed, market-based instruments are already in operation in different parts of India. For example, an arrangement to share benefits (an in-kind payment) among villagers who participated in protecting upstream water sources from siltation in the Sukhomajri watershed region in Northern India is a classic example of how PES-like programmes can work efficiently in the Indian context (Huang et al., 2009; Kerr, 2002). Similarly, in Maharashtra, unused water from reservoirs of incomplete irrigation projects is being successfully transported, with the help of private operators and through networked pipes (mostly underground), to supply water to the farmers. There are around 100,000 such schemes successfully operating in Maharashtra (Ackermann, 2013).

In Tamil Nadu, water users use market-based instruments to manage irrigation tanks and to allocate irrigation services in an efficient and equitable manner. For example, the Rettaikulam tank, in Thirunelveli District of Tamil Nadu, exemplifies an efficiently functioning water tax system (Sakthivadivel et al., 2004). This example shows that the user groups managing the tanks levy *ayacut vari* (a tax based on landholding) and use the tax revenue to meet the financial requirements for maintaining the tanks. The tax rate per acre is determined by the user groups, based on the extent of repair and maintenance work to be done and the funds required for such work. The tax is collected from water users in the command area.

Some empirical studies demonstrate that farmers in certain river basins in South India are willing to trade their excess water to other farmers, provided they are adequately compensated (Biswas, 2015; Venkatachalam & Narayanamoorthy, 2012). A similar kind of trade can take place between farmers and high-value users of water, as the compensation in this case may be much higher than that of water trade among farmers themselves. Such practices are based on the *user-pays* principle, a fundamental principle of market-based instruments, which provides financial self-sufficiency and incentives for the user groups to sustain conservation efforts collectively. Since the user groups act as utility-maximizing individuals, introducing PES within an appropriate institutional set-up (Asquith & Wunder, 2008) generates adequate economic incentives for conservation of critical resources as well as further scaling-up. Experience with PES schemes around the world suggests that by implementing PES effectively, future water transfer from irrigation tanks to high-value urban use could generate a win-win outcome for water managers, farmers, municipalities and urban dwellers.

Conditions for implementing water PES

Despite many advantages, PES schemes face many implementation challenges (Behera et al., 2011). The very fact that PES has not yet percolated deeply in the environmental domain suggests that many details need to be worked out during implementation. It may work better under certain conditions, such as in conducive ecological settings, with improved nature and quantity of services and their continuous provision, well-defined and secure property rights, appropriate legal framework, and trust among the parties involved (Behera et al., 2011). Broader guidelines for success have emerged from several PES schemes implemented in other parts of the world, such as Vietnam, Indonesia, China and Mexico (Adhikari, 2009; Alix-Garcia et al., 2008; Asquith & Wunder, 2008; Huang et al., 2009; Pagiola, 2002).

The prerequisites for water PES to work efficiently are: water accounting in physical units wherever possible (Perry, 2013); economic valuation of water and its ecosystem services; and institutional arrangements for facilitating voluntary trade of water services. Before implementing water PES for tanks, accounting for water resources – in physical as well as in economic terms – will have to be systematically done. In the case of system tanks (fed by a river system), river-basin-level water accounting will be more appropriate, as the hydrological changes taking place in the entire river basin have profound impacts on the water dynamics of the system tanks. In the case of physical accounting for water, both the 'stock' and the 'flow' components of water resources get into the accounting matrices. While the stock account considers the stock of water resources in the opening and closing periods, the flow account captures the 'additions' that increase the level of stock (total precipitation, inflow from tributaries, return flow from use sectors, and import from other basins) and 'subtractions' that reduce the stock (evaporation, evapotranspiration, non-recoverable deep percolation, amount of water consumed for various economic and non-economic entities, water exported to other basins, and water drained into the ocean). The net change (surplus or deficit) in the stock of the water in physical terms between accounting periods can be arrived at from the stock account. The flow account, on the other hand, depicts what happens to the available water in the basin, how much water is used for productive purposes, where the unproductive water goes and how much surplus water is available for transfer. Though 'physical accounts' are a necessary condition for water allocation decisions, 'economic accounts' fulfil the sufficient condition since allocation decisions will have to be based largely on the economic value of scarce water.

Economic accounts try to place a monetary value not only on the net change in the stock but also on different levels of services and benefits from water. It helps in assessing the marginal efficiency of water used in a particular sector as well as the marginal gains and losses of allocating water from a less productive to a more productive use or from a less efficient to a more efficient use (Balooni & Venkatachalam, 2016). The monetary value of water can be estimated by using non-market valuation techniques. Alternatively, such values can be generated by using a *benefit transfer method* (Plummer, 2009) – a method by which economic values of ecosystem services are derived from previous economic valuation studies – as the economic values for water-related ecosystem services have been already estimated by a significant number of non-market valuation studies in the Indian context (Kumar, Vedantam, et al., 2012; Mukherjee & Kumar, 2012). For non-system tanks, water accounting at a tank level will yield fruitful results.

Institutional arrangements for an efficient PES scheme and its sustainability depend largely on how different institutions – formal and informal, external and internal, modern and traditional – collaborate to enhance trade in environmental services (Greiber, 2009; Vatn, 2009). However, an unmet challenge is to identify an appropriate combination of different institutions that can produce economically efficient outcomes. Path dependency guides us when our decisions are governed by bounded rationality, i.e. efficient decisions are usually constrained by our lack of ability to collect and process relevant information, which in turn is influenced by our limited cognitive abilities. Drawing lessons from experience, an outline of the additional institutional arrangements required for PES schemes is laid down here.

First of all, identifying the sellers and buyers of water and assigning property rights over water in the irrigation tanks are prerequisites for effective implementation of PES schemes. In the case of tank water, the municipalities are the buyers, and the farmers are the sellers. If the farmers have to change their land-use patterns in various tank commands to save water for urban use, then WUAs can be assigned the property rights over the tank, and thus they become the owners of the water rights. The WUAs can in turn assign water rights to individual farmers. The farmers can be incentivized by the WUAs to use water-efficient crops and efficient irrigation technologies so that they can transfer the saved water to the WUAs (for a price), which in turn can sell the rights to the urban utility. Though WUAs have been created in different states, the underlying incentive/disincentive structure for their efficient functioning is not well defined. Water PES can make WUAs work more efficiently without additional transaction costs, since water PES would be built on the already existing WUAs. As the resource to be managed is relatively small, the WUAs have comparative advantages in monitoring and regulating water use (Fisher et al., 2010), curtailing 'free-riders'. If WUAs do not exist, or if the irrigation tanks belong to the *panchayat*, (i.e. the local government elected by the people to manage public goods such as irrigation canals and grazing lands, and deliver services such as water supply), then the property rights over the tank water could be assigned to the panchayat, and therefore the panchayat will have the selling rights; the households should unanimously decide on how to use the sales revenue through negotiations (Turpie et al., 2008). If the ecosystem services are sizeable, then the panchayats may be assigned property rights, and the WUAs/farmer groups can be part of the larger stakeholders making decisions.

Water PES works well if: (a) the buyers use the water for higher-value uses than the sellers; (b) the opportunity cost of providing water services is less than or equal to the amount paid to the service providers (Kosoy, Martinez-Tuna, Muradian, & Martinez-Alier, 2007) – in other words, the scheme can work smoothly only when willingness to pay is greater than or equal to willingness to accept compensation; and (c) transaction costs are negligible (Tacconi, 2012). In the Indian context, Venkatachalam (2015) has empirically demonstrated that urban consumers are already spending a significant part of their income on water and are willing to pay more for improved water supply.

PES can generate additional income for local farmers and can contribute to alleviating poverty (Tang, Shi, Nan, & Xu, 2012). For example, it is largely the poor who depend on the irrigation tanks for their livelihoods (Balasubramanian & Selvaraj, 2003; Narayanamoorthy, 2007), and the share of the economic benefits from ecosystem services from the tanks in the total income of the poor households is greater than that of the rich. So, water PES can generate relatively larger benefits for poor people. If the minimum number of stakeholders is not available for maintaining the tanks, then tank management becomes problematic because the higher marginal cost of maintenance will have to be borne by a smaller number of water users. On the other hand, some rich farmers may switch from tank water to tube-wells, which would make more water available to the remaining farmers who happen to be poor. Poor farmers will have an incentive to efficiently manage the tanks if their expected benefits from their collective management efforts are substantial (Kumar, Bassi, Kishan, Chattopadhyay, & Ganguly, 2016). Hence, PES can not only allocate water more efficiently but can also contribute to local poverty reduction.

Conclusions

When agricultural land is acquired for non-agricultural purposes, the water which was used for irrigation in that land is lost. When water is withdrawn for urban use, the issues involved are not adequately addressed in the water development literature, especially in the Indian context. The wider supply–demand gap in urban water supply forces governments to withdraw water from distant irrigation sources, as is evident in the case of Tamil Nadu in India. In most cases, the irrigation sources happen to be small water bodies like irrigation tanks and lakes, which provide valuable ecosystem services to the local people.

Withdrawal of water from water bodies that are traditionally used for irrigation is done by government agencies, and as a result, the farmers depending on those water sources become the net losers. The trade-off involved in transferring water from agriculture to urban use can be reduced and a win-win situation can be created if innovative and efficient institutions are introduced for water transfer. This study described how a market-based approach, that is, water PES, can be effectively employed to make water reallocation beneficial to both the farmers and the urban consumers. Though water PES is relatively more efficient than the current command-and-control method of water transfer, the effectiveness of operationalizing the concept depends largely on institutional arrangements that appropriately combine the role of market, government, non-governmental organizations and user groups (Behera et al., 2011).

This study emphasizes that either the WUAs or the panchayat system managing the irrigation systems at the local level should own the water rights, so that the benefits of managing and transferring water from the tanks are fairly distributed among all those involved in tank management. However, proper water accounting in both physical and economic terms is key to making water trading efficient. A word of caution, though: there are no standard panaceas, and all negative externalities cannot be internalized with a typical model (Ostrom, Janssen, & Anderies, 2007). The effectiveness of PES depends mainly on how complexities related to uncertainty, distributional issues, social harmony and power relations, prevailing especially at the regional and local levels, are taken into account while designing such schemes (Muradian, Corbera, Pascual, Kosoy, & May, 2010).

To conclude, employment of PES at the grass-roots level to cater to the demand of the bourgeoning population in urban areas has the potential to produce efficient outcomes, as this study has shown in the context of irrigation tanks in the state of Tamil Nadu. However, there is a need to tread cautiously, as this institutional model needs further investigation.

Acknowledgements

The authors sincerely thank A. Narayanamoorthy for encouraging us to undertake this study. The authors are also thankful to M. Dinesh Kumar, David Zetland, Bhagirath Behera, Mrinal K. Dutta, Durba Biswas and Chandra Sekhar Bahinipati for valuable comments on an earlier version of this article.

Disclosure statement

No potential conflict of interest was reported by the authors.

References

Ackermann, R. (2013). Improving water use efficiency: New directions for water management in India. In M. Ferroni (Ed.), *Transforming Indian agriculture India 2040: Productivity, markets, and institutions* (pp. 127–185). New Delhi: Sage.

Adhikari, B. (2009). *Market-based approaches to environmental management: A review of lessons from payment for environmental services in Asia* (ADBI Working Paper Series No. 134). Retrieved December 10, 2016, from http://www.adbi.org/files/2009.03.26.wp134.market.based.approaches.environmental.mngt.pdf

Alix-Garcia, J., de Janvry, A., Sadoulet, E., & Torres, J. M. (2008). Lessons learned from Mexico's payment for environmental services program. In L. Lipper, T. Sakuyama, R. Stringer, & D. Zilberman (Eds.), *Payment for environmental services in agricultural landscapes: Economic policies and poverty reduction in developing countries* (pp. 163–188). Rome, London: FAO, Springer.

Asquith, N., & Wunder, S. (2008). *Payments for watershed services: The Bellagio conversations*. Santa Cruz de la Sier: Fundacion Natura Boliviara.

Balasubramanian, R., & Selvaraj, K. N. (2003). *Poverty, private property and common pool resource management: The case of irrigation tanks in south India* (SANDEE Working Paper No. 2-03). Kathmandu: SANDEE.

Balooni, K., & Venkatachalam, L. (2016). Managing water for sustainable development: An Indian perspective. *IIM Kozhikode Society and Management Review, 5*, vii–xiii. doi:10.1177/2277975215625500

Barbier, E. B., Koch, E. W., Silliman, B. R., Hacker, S. D., Wolanski, E., & Reed, D. J. (2008). Coastal ecosystem-based management with nonlinear ecological functions and values. *Science, 319*, 321–323. doi:10.1126/science.1150349

Behera, B., Mishra, P., & Nayak, N. C. (2011). 'Payments for environmental services: Issues and implications for India. *Economic and Political Weekly, 50*, 64–68.

Bhatt, S. (2013). How does participatory irrigation management work? A study of selected water users' associations in Anand district of Gujarat, western India. *Water Policy, 15*, 223–242. doi:10.2166/wp.2012.065

Biswas, D. (2015). *Farmers' willingness to pay for improved irrigation water: A case study of Malaprabha irrigation project in Karnataka, India*. Paper presented in International Seminar on Natural Resources and National Accounts in South Asia, Bangalore, February 5–6, Institute for Social and Economic Change.

Chandran, R. (2016). Bengaluru makes a splash with Cauvery water as farmers in nearby villages suffer. *Livemint*, E-Paper. Retreived December 12, 2016, from http://www.livemint.com/Politics/tvTkH9u8fFL4HcA3RFfH8 K/Bengaluru-makes-a-splash-with-Cauvery-water-as-farmers-in-ad.html

Chaudhry, S. (2015). Land acquisition bill implies deep trouble. *DNA*, Mumbai. Retrieved December 12, 2016, from http://www.dnaindia.com/india/report-land-acquisition-bill-implies-deep-trouble-2072222

Clements, T., John, A., Nielsen, K., An, D., Tan, S., & Milner-Gulland, E. J. (2010). Payments for biodiversity conservation in the context of weak institutions: Comparison of three programs from Cambodia. *Ecological Economics, 69*, 1283–1291. doi:10.1016/j.ecolecon.2009.11.010

Cullet, P., Bhullar, L., & Koonan, S. (2015). Inter-sectoral water allocation and conflicts: Perspectives from Rajasthan. *Economic and Political Weekly, 50*, 61–69.

Engel, S., Pagiola, S., & Wunder, S. (2008). Designing payments for environmental services in theory and practice: An overview of the issues. *Ecological Economics, 65*, 663–674. doi:10.1016/j.ecolecon.2008.03.011

Fisher, B., Kulindwa, K., Mwanyoka, I., Turner, R. K., & Burgess, N. D. (2010). Common pool resource management and PES: Lessons and constraints for water PES in Tanzania. *Ecological Economics, 69*, 1253–1261. doi:10.1016/j.ecolecon.2009.11.008

Gandy, M. (2008). Landscapes of disaster: Water, modernity, and urban fragmentation in Mumbai. *Environment and Planning A, 40*, 108–130. doi:10.1068/a3994

Ghatak, M., Mitra, S., Mookherjee, D., & Nath, A. (2013). Land acquisition and compensation: What really happened in Singur? *Economic and Political Weekly, 48*, 32–44.

Ghosh, B. (2012). What made the 'unwilling farmers' unwilling? A note on Singur. *Economic and Political Weekly, 47*, 13–16.

Greiber, T. (2009). *Payments for ecosystem services: Legal and institutional frameworks* (IUCN Environmental Policy and Law Paper No. 78). Switzerland: IUCN.

Huang, M., Upadhyaya, S. K., Jindal, R., & Kerr, J. (2009). Payments for watershed services in Asia: A review of current initiatives. *Journal of Sustainable Forestry, 28*, 1–25.

Jegadeesan, M., & Koiji, F. (2011). Deterioration of the informal tank institution in Tamil Nadu: Caste-based rural society and rapid economic development in India. *Southeast Asian Studies, 49*, 93–123.

Kerr, J. (2002). Sharing the benefits of watershed management in Sukhomajri, India. In S. Pagiola, J. Bishop, & N. Landell-Mills (Eds.), *Selling forest environmental services: Market-based mechanisms for conservation and development* (pp. 63–75). London: Earthscan.

Kolstad, C. D. (2011). *Intermidtae environmental economics* (2nd ed.). Oxford: Oxford University Press.

Kosoy, N., Martinez-Tuna, M., Muradian, R., & Martinez-Alier, J. (2007). Payments for environmental services in watersheds: Insights from a comparative study of three cases in Central America. *Ecological Economics, 61*, 446–455. doi:10.1016/j.ecolecon.2006.03.016

Kumar, D. M., Bassi, N., Kishan, K. S., Chattopadhyay, S., & Ganguly, A. (2016). Rejunevating Tanks in Telengana. *Economic and Political Weekly, 51*, 30–34.

Kumar, D. M., & Rao, N. (2017). The hydro-institutional challenge of managing tanks: A study of tanks in rural Andhra Pradesh. In H. Sandhu (Ed.), *Ecosystem functions and management* (pp. 155–191). Switzerland: Springer. doi:10.1007/978-3-319-53967-6

Kumar, D. M., Vedantam, N., Bassi, N., Puri, S., & Sivamohan, M. V. K. (2012). *Making rehabilitation' work: Protocols for improving performance of irrigation tanks in Andhra Pradesh*. Hyderabad: Institute for Resource Analysis and Policy.

Kumar, M. D. (2014). *Thirsty cities: How Indian cities can meet their water needs*. New Delhi: Oxford University Press.

Kumar, M. D., Panda, R., Niranjan, V., & Bassi, N. (2012). Technology choices and institutions for improving economic and livelihood benefits from multiple uses tanks in western Orissa. In M. D. Kumar, M. V. K. Sivamohan, & N. Bassi (Eds.), *Water management, food security and sustainable agriculture in developing economies* (pp. 138–163). London: Routledge/Earthscan.

Lakshmi, K. (2013). Rejuvenate Chennai tanks: Study. *The Hindu*. Retrieved December 11, 2016, from http://www.thehindu.com/news/cities/chennai/rejuvenate-chennai-tanks-study/article5097814.ece

Lele, S., Patil, I., Badiger, S., Menon, A., & Kumar, R. (2011). Forests, hydrological services, and agricultural income: A case study from Mysore district of the Western Ghats of India. In A. K. E. Haque, M. N. Murty, & P. Shyamsundar (Eds.), *Environmental Valuation in South Asia* (pp. 141–169). Cambridge: Cambridge University Press.

Levien, M. (2012). The land question: Special economic zones and the political economy of dispossession in India. *The Journal of Peasant Studies, 39*, 933–969. doi:10.1080/03066150.2012.656268

Lipper, L., Sakuyama, T., Stringer, R., & Zilberman, D. (2009). *Payment for environmental services in agricultural landscapes*. London: Springer.

Locatelli, B., Rojas, V., & Salinas, Z. (2008). Impacts of payments for environmental services on local development in northern Costa Rica: A fuzzy multi-criteria analysis. *Forest Policy and Economics, 10*, 275–285. doi:10.1016/j.forpol.2007.11.007

Meinzen-Dick, R. (2007). Beyond panaceas in water institutions. *Proceedings of the National Academy of Sciences, 104*, 15200–15205. doi:10.1073/pnas.0702296104

Millennium Ecosystem Assessment. (2005). *Ecosystems and human well-being: Synthesis*. Washington, DC: Island Press.

MoEF. (2006). *National environment policy 2006*. New Delhi: Ministry of Environment and Forests. Retrieved December 11, 2016, from http://moef.nic.in/downloads/about-the-ministry/introduction-nep2006e.pdf.

Molle, F., & Berkoff, J. (2006). *Cities versus agriculture: Revisiting intersectoral water transfers, potential gains and conflicts* Comprehensive Assessment Research Report 10). Colombo: International Water Management Institute.

MoWR. (2012). *National water policy, 2012*. New Delhi: Ministry of Water Resources. Retrieved December 11, 2016, from http://wrmin.nic.in/writereaddata/NationalWaterPolicy/NWP2012Eng6495132651.pdf

Mukherjee, S., & Kumar, M. D. (2012). Economic valuation of a multiple use wetland water system: A case study from India. *Water Policy, 14*, 80–98. doi:10.2166/wp.2011.120

Mukherji, A., Blanka, F., Shah, T., Suhardiman, D., Giordano, M., & Weligamage, P. (2010). *Irrigation reform in Asia: A review of 108 cases of irrigation management transfer* (IWMI Research Report). Colombo: International Water Management.

Munoz-Pina, C., Guevara, A., Torres, J. M., & Brana, J. (2008). Paying for the hydrological services of Mexico's forests: Analysis, negotiations and results. *Ecological Economics, 65*, 725–736. doi:10.1016/j.ecolecon.2007.07.031

Muradian, R., Corbera, E., Pascual, U., Kosoy, N., & May, P. H. (2010). Reconciling theory and practice: An alternative conceptual framework for understanding payments for environmental services. *Ecological Economics, 69*, 1202–1208. doi:10.1016/j.ecolecon.2009.11.006

Nalukenge, I., Antle, J. M., & Stoorvogel, J. (2008). Assessing the feasibility of wetlands conservation: Using payment for ecosystem services in Pallisa, Uganda. In L. Lipper, T. Sakuyama, R. Stringer, & D. Zilberman (Eds.), *Payment for environmental services in agricultural landscapes: Economic policies and poverty reduction in developing countries* (pp. 239–253). Rome, London: FAO, Springer.

Narain, V. (2009). Growing city, shrinking hinterland: Land acquisition, transition and conflict in peri-urban Gurgaon. *Environment and Urbanization, 21*, 501–512. doi:10.1177/0956247809339660

Narayanamoorthy, A. (2004). Impact assessment of drip irrigation in India: The case of sugarcane. *Development Policy Review, 22*, 443–462. doi:10.1111/dpr.2004.22.issue-4

Narayanamoorthy, A. (2007). Tank irrigation in India: A time series analysis. *Water Policy, 9*, 193–216. doi:10.2166/wp.2006.063

Narayanamoorthy, A., & Suresh, R. (2017). Does urbanisation affect tank irrigation development in Tamil Nadu? *A Macro-Level Analysis, Review of Development and Change, 21*, 9–32.

North, D. C. (1990). *Institutions, institutional change and economic performance*. Cambridge, MA: Cambridge University Press.

Ostrom, E., Janssen, M. A., & Anderies, J. M. (2007). Going beyond panaceas. *Proceedings of the National Academy of Sciences, 104*, 15176–15178. doi:10.1073/pnas.0701886104

Pagiola, S. (2002). Paying for water services in central America: Learning from Costa Rica. In S. Pagiola, J. Bishop, & N. Landell-Mills. (Eds.), *Selling Forest environmental services: Market-based mechanisms for conservation and development* (pp. 37–62). London: Earthscan.

Pagiola, S. (2008). Payments for environmental services in Costa Rica. *Ecological Economics, 65*, 712–724. doi:10.1016/j.ecolecon.2007.07.033

Palanisami, K., Meinzen-Dick, R., & Giordano, M. (2010). Climate change and water supplies: Options for sustaining tank irrigation potential in India. *Economic and Political Weekly, 45*, 183–190.

Pant, N. (2008). Some issues in participatory irrigation management. *Economic and Political Weekly, 43*, 30–36.

Perrot-Maitre, D. (2006). *The Vittel payments for ecosystem services: A "perfect" PES case?* London: International Institute for Environment and Development.

Perry, C. (2013). 'ABCDE+F: A framework for thinking about water resources management. *Water International, 38*, 95–107. doi:10.1080/02508060.2013.754618

Plummer, M. L. (2009). Assessing benefit transfer for the valuation of ecosystem services. *Frontiers in Ecology and the Environment, 7*, 38–45. doi:10.1890/080091

Ramachandraiah, C., & Srinivasan, R. (2011). Special economic zones as new forms of corporate land grab: Experiences from India. *Development, 54*, 59–63. doi:10.1057/dev.2010.99

Reddy, N. D., & Mishra, S. (Eds.). (2009). *Agrarian crisis in India*. New Delhi: Oxford University Press.

Reddy, P., & Reddy, R. (2005). How participatory is participatory irrigation management? *Economic and Political Weekly, 60*, 5587–5595.

Sakthivadivel, K., Gomathinayagam, P., & Shah, T. (2004). Rejuvenating irrigation tanks through local institutions. *Economic and Political Weekly, 34*, 3521–3526.

Schomers, S., & Matzdorf, B. (2013). Payments for ecosystem services: A review and comparison of developing and industrialized countries. *Ecosystem Services, 6*, 16–30. doi:10.1016/j.ecoser.2013.01.002

Seenivasan, R., & Kanagavalli, J. (2014). Dying tanks in urban areas: What can be done with them? *Review of Development and Change, 19*, 109–122.

Shah, T., & Raju, K. V. (2001). *Rethinking rehabilitation: Socio-ecology of tanks and water harvesting in Rajasthan, north-west India* (CAPRi Working Paper No. 18). Washington, DC: IFPRI.

Singh, R. (2012). Inefficiency and abuse of compulsory land acquisition. *Economic and Political Weekly, 47*, 46–53.

Tacconi, L. (2012). Redefining payments for environmental services. *Ecological Economics, 73*, 29–36. doi:10.1016/j.ecolecon.2011.09.028

Tang, Z., Shi, Y., Nan, Z., & Xu, Z. (2012). The economic potential of payments for ecosystem services in water conservation: A case study in the upper reaches of Shiyang River basin, northwest China. *Environment and Development Economics, 17*, 445–460. doi:10.1017/S1355770X12000149

TEEB. (2010). The economics of ecosystems and biodiversity: Mainstreaming the economics of nature: A synthesis of the approach, conclusions and recommendations of TEEB. Retrieved December 10, 2016, from www.teebweb.org/wp-content/uploads/Study%20and%20Reports/Reports/Synthesis%20report/TEEB%20Synthesis%20Report%202010.pdf

Turpie, J. K., Marais, C., & Blignaut, J. N. (2008). The working for water programme: Evolution of a payments for ecosystem services mechanism that addresses both poverty and ecosystem service delivery in South Africa. *Ecological Economics, 65*, 788–798. doi:10.1016/j.ecolecon.2007.12.024

Vatn, A. (2009). An institutional analysis of payment for environmental services. *Ecological Economics, 69*, 1245–1252.

Venkatachalam, L. (2015). Informal water markets and willingness to pay for water: A case study of urban poor in Chennai city, India. *International Journal of Water Resources Development, 31*, 134–145. doi:10.1080/07900627.2014.920680

Venkatachalam, L., & Jayanthi, M. (2016). Economic valuation of wetland ecosystem services: A contingent valuation approach. *Review of Development and Change, 21*, 89–110.

Venkatachalam, L., & Narayanamoorthy, A. (2012). Estimating economic value of irrigation water through contingent valuation method: Results from Bhavani river basin, Tamil Nadu. *Indian Journal of Agricultural Economics, 67*, 308–315.

Vij, S., & Narain, V. (2016). Land, water & power: The demise of common property resources in periurban Gurgaon, India. *Land Use Policy, 50*, 59–66. doi:10.1016/j.landusepol.2015.08.030

Wunder, S. (2008). Payments for environmental services and the poor: Concepts and preliminary evidence. *Environment and Development Economics, 13*, 279–297.

The negative impact of subsidies on the adoption of drip irrigation in India: evidence from Madhya Pradesh

R. P. S. Malik, Mark Giordano and M. S. Rathore

ABSTRACT
Drip irrigation in India has expanded slowly. One reason cited is the high capital costs facing the smallholder-dominated agricultural sector. Governments have provided capital subsidies in response. This study finds that, rather than improving access to drip, the subsidy system holds the technology back, because its technical requirements, highly bureaucratic processes and pricing incentives turn many drip providers into rent-seeking agents rather than service providers to farmers, leading to price increases of 40% or more. If capital costs are truly the constraint on drip expansion in India, alternative models to address them are available.

Introduction

Growing water scarcity has forced many countries to look for opportunities to increase water savings and productivity, including in the agricultural sector, where most water use takes place. Given the difficulties in implementing demand-management schemes, especially in developing countries, where the number of users is large and implementation of levers such as water pricing is difficult, efforts tend to focus on technology-based, supply-side solutions. One such technology receiving considerable attention is drip irrigation.

Drip has many potential advantages, and a large literature in fact suggests that adoption of drip irrigation technologies conserves water, improves crop yields and increases water and fertilizer use efficiency and that investments in drip are cost-effective, with short payback periods (see e.g. the review by van der Kooij, Zwarteveen, Boesveld, & Kuper, 2013). Nonetheless, adoption of drip irrigation by farmers often disappoints its promoters. Globally, drip still accounts for less than 5% of irrigated area (International Commission on Irrigation & Drainage, 2012). In India estimates of potential drip-irrigated area range from around 12 million hectares (Raman 2010; Palanisami, Mohan, Kakumanu, & Raman, 2011) to more than 25 million hectares (Government of India, 2004). Even using the lower, more conservative (Bhamoriya & Mathew, 2014) figure, Palanisami et al. estimate that only around 13% of the potential area is now covered by drip irrigation. Venot et al. (2014) highlight that while reduction in groundwater overdraft is a key reason for drip promotion, the drip-irrigated area is equivalent to only 5% of India's groundwater-irrigated area. One reason cited to explain the gap between the high potential and the low uptake is the high initial investment costs for drip, especially for the small farmers who make up the majority of India's agricultural sector (Narayanamoorthy, 2009).

In response, the government of India, in conjunction with state governments, has provided capital-cost subsidies, ranging over time from 30% to 90% of purchase costs, for potential drip users (Narayanamoorthy, 2012). Demonstration effects by the early adopters benefitting from the subsidies are expected to further encourage adoption. The overall increase in demand is then expected to spur innovation in the drip industry, reduce production costs and drive down prices, further increasing the use of drip irrigation technology and increasing India's overall water-use efficiency. However, evidence suggests that, if anything, the subsidy system may actually constrain the rate of drip adoption and drive prices up.

This article examines the question of whether and how the drip subsidy system in India impacts adoption. It is primarily based on data and information collected through extensive personal interviews with stakeholders involved in the use, manufacture, retail, promotion (NGOs, extension agencies) and administration of drip-technology programmes in Madhya Pradesh, the state calculated to have the third-largest potential drip area in India (Raman, 2010). Interviews were designed to provide insights into how the subsidy system is conceived, how it works in practice, and its likely impact on drip adoption.

The article finds that rather than promoting drip irrigation, subsidies hold the technology back. This is because the well-meaning but convoluted nature of the subsidy delivery system, with its specific technical requirements, highly bureaucratic processes and pricing incentives, turns many drip providers into agents seeking revenues from the government instead of providing service to farmers. As a result, costs appear to be higher than they would be without the subsidy in place. The overall system has also made manufacturers subservient to government favours, reduced enterprise spirit and encouraged poor business practices. These findings are of course damning for the current drip subsidy system in Madhya Pradesh, but they also provide insights into how subsidy schemes for low-income farmers could be better developed in that state, elsewhere in India and perhaps further afield.

Drip irrigation and its use in India

Drip irrigation has the potential to increase water-use efficiency by slowly dripping water from small holes in plastic lines to plant roots, either directly or from the soil surface. The precise application reduces evaporative losses, as well as losses through deep percolation to the groundwater table. Drip irrigation technologies range from simple bucket-kit systems for small farms to automated systems linking release of water to soil moisture conditions measured continuously by tensiometers.

While drip-like technologies have been in use since ancient times, it has only been with the advent of modern plastics that major improvements in drip irrigation have become possible. Water in a modern drip system is delivered through a network of valves, pipes, tubing and emitters. Most large drip irrigation systems employ screen filters to prevent clogging of emitters by small waterborne particles and permit delivery of liquid fertilizers and other nutrients along with the water, in a process known as fertigation or chemigation, using injectors such as piston or Venturi pumps. Using drip, water and fertilizer use efficiency are often said to have the potential to reach or exceed 90%. In addition, drip irrigation is reported to reduce problems of salinization and waterlogging and in water-scarce environments may allow agriculture where furrow or flood irrigation would not be possible.

As awareness of water scarcity in India has increased over the past few decades, there have been growing calls to use technologies such as drip to increase agricultural water-use efficiency. The use of drip irrigation in India began with initial testing at Tamil Nadu Agricultural University in Coimbatore in 1970, and the area under drip had reached 55,000 ha by 1992 (Polak and Sivanapan, 1998). The technology was introduced on a commercial scale during the Eighth Five-Year Plan (1992–97), and by 2003 the area reported under drip had grown tenfold, to 0.5 million ha (GOI, 2004). However, this figure was still only 2–5% (Raman, 2010; INCID, 1994) of the 27 million ha (GOI, 2004) of land officially estimated to be suitable for drip – far short of government expectations, especially given the many apparent benefits. In response to low adoption, a task force on micro-irrigation was set up by the government of India to determine how to further encourage drip use, and targets of 12 million ha under drip irrigation by 2012, and the entire drip-suitable area under drip by 2030, were set (GOI, 2004).

Drip subsidies in India and Madhya Pradesh

One key constraint identified by the task force as a hindrance to adoption was the high initial investment costs for farmers. The task force and subsequent Five-Year Plan thus focused on capital subsidies as the primary vehicle for promoting drip expansion. However, capital subsidies were not new to drip promotion in India. The first capital subsidies were introduced by the state of Maharashtra in 1988 and followed in other states. The first centrally sponsored scheme on the use of plastics in agriculture was launched during the Eighth Five-Year Plan (1992–97) with an outlay of INR 2500 million, of which four-fifths was earmarked to promote drip and other micro-irrigation. Under the national system, the central government paid the majority of subsidy costs, with state governments covering the remainder. The schemes had been modified over time in terms of the total value of the subsidy, the proportion of drip costs subsidized, the institutional arrangements for disbursal of subsidy, and the eligibility criterion and other conditionalities attached to the subsidies.

Following the recommendations of the 2004 task force, the government of India launched a National Mission on Micro Irrigation as a new centrally sponsored scheme to further promote drip and other micro-irrigation approaches through financial subsidies. Components of drip and sprinkler irrigation systems were also included in other centrally sponsored schemes, including the National Food Security Mission; Integrated Scheme of Oilseeds, Pulses, Oil Palm and Maize; and the Technology Mission on Cotton. These programmes conformed to the same norms and pattern of assistance as stipulated under the National Mission on Micro Irrigation. Under all the programmes, 40% of the cost of the drip irrigation system was to be borne by the central government, 10% by the relevant state government and the remaining amount by the beneficiary, either through his/her own resources or through loans from financial institutions. An additional 10% subsidy was given for recipients classified as small or marginal farmers and belonging to Scheduled Castes or Scheduled Tribes, though some states provide amounts beyond this requirement. In the specific case of Madhya Pradesh, total drip subsidies are 70–80% of the approved cost of a drip system, as shown in Table 1.

Table 1. Drip irrigation subsidies in Madhya Pradesh (2011).

Farmer and caste classification	Percent subsidy		
	Central government	State government	Total
Small and marginal farmers of Scheduled Castes and Tribes*	50	30	80
Other farmers of Scheduled Castes and Tribes*	40	30	70
All other small and marginal farmers	50	20	70
All other farmers	40	30	70

*Scheduled Castes and Tribes refer to various groups of historically socially disadvantaged people in India. The terms are recognized in the Constitution of India.
Source: Government of Madhya Pradesh.

Subsidy practice in Madhya Pradesh

It has been estimated that Madhya Pradesh has 1.38 million ha of land suitable for coverage with drip irrigation (Raman, 2010). While drip irrigation statistics are not systematically collected in India, official data show that as of 2008, Madhya Pradesh's total drip area was 20,432 ha, or about 1% of the potential (Indiastat). The failure to cover more area is not due to lack of suppliers. There are at least 22 companies operating in Madhya Pradesh providing drip and sprinkler irrigation systems, and another 6 providing only drip. Some, such as Jain Irrigation and Netafim Irrigation, are recognized for the high quality of their products. Some others are considered lower-quality suppliers, and yet others fall in the middle. About one-quarter of these companies are based in Madhya Pradesh.

What is the basis for subsidy eligibility and calculation?

To ensure that only high-quality drip products are subsidized by the government, potential supplier participants in the drip subsidy programme must obtain certification that their products conform to standards stipulated by the Bureau of Indian Standards. Manufacturers of higher-quality drip products obtain the certificate by bringing their products in for inspection. Interviews revealed that other manufacturers may pay 'convenience' fees to avoid physical inspection and still obtain the certificate.

The government of India has also established detailed specifications for what they define as typical drip systems to help ensure that certified companies create packages for farmers which can be turned into a fully functional drip system. Component specifications for two of these systems are given in Table 2 as examples and to highlight the level of detail in the specifications. Subsidies apply only when the set of components is purchased as a whole and are not available for individual components of the system. Non-governmental organizations involved in drip supply, such as international development enterprises, and private companies focused on low-cost/low-quality systems aimed at poor farmers, are not eligible for subsidies.

Based on specifications such as these, the National Mission on Micro Irrigation estimates drip system costs for particular farming system requirements as shown in Table 3. The share of these costs eligible for subsidy is based on the criteria described in Table 1, irrespective of whether the prices of a specific certified manufacturer are higher or lower than those indicated.

Table 2. Detailed component and material requirements for two drip irrigation systems eligible for subsidy.

Wide-spaced crops, 0.2 ha	Wide-spaced crops, 5 ha
PVC pipe 75 mm, Class II, 4 kg/cm^2	PVC pipe 90 mm, Class II, 4 kg/cm^2
PVC pipe 63 mm, Class II, 4 kg/cm^2	PVC pipe 75 mm, Class II, 4 kg/cm^2
PVC pipe 50 mm, Class II, 4 kg/cm^2	PVC pipe 63 mm, Class II, 4 kg/cm^2
Lateral 16 mm, Class II, 2.5 kg/cm^2	Lateral 16 mm, Class II, 2.5 kg/cm^2
Lateral 12 mm, Class II, 2.5 kg/cm^2	Lateral 12 mm, Class II, 2.5 kg/cm^2
Emitter 4–8 L/h	Emitter 4–8 L/h
Microtube 6 mm	Microtube 6 mm
Control valve 75 mm	Control valve 90 mm
Control valve 63 mm	Control valve 75 mm
Control valve 50 mm	Control valve 63 mm
Flush valve 63 mm	Flush valve 75 mm
Flush valve 50 mm	Flush valve 63 mm
Air release valve 1″	Air release valve 1.5″
Non-return valve 1.5″	Non-return valve 1.5″
Throttle valve 1.5″	Non-return valve 2.5″
Screen filter 10 m^3/h	Throttle valve 1.5″
Bypass assembly 2″	Throttle valve 2″
Bypass assembly 1.5″	Throttle valve 2.5″
Venturi & manifold 2″	Screen filter 20–25 m^3/h
Venturi & manifold 1.5″	Screen filter 10 m^3/h
	Bypass assembly 2.5″
	Bypass assembly 2″
	Bypass assembly 1.5″
	Venturi & manifold 2.5″
	Venturi & manifold 2″

Source: Government of India (2010).

Table 3. Indicative costs (in INR) of drip irrigation systems for subsidy calculation based on farm size and lateral spacing (2010).

		Farm size						
Lateral spacing (m)	Spacing	0.2 ha	0.4 ha	1.0 ha	2.0 ha	3.0 ha	4.0 ha	5.0 ha
12 × 12	Wide	8,057	13,785	18,820	29,928	46,467	57,809	73,611
8 × 8	Wide	8,673	15,088	22,028	36,217	56,087	70,893	89,964
4 × 4	Wide	11,177	18,621	31,793	55,725	86,926	113,812	135,459
2 × 2	Wide	18,319	31,616	63,598	123,441	179,332	249,134	305,797
1.2 × 0.6	Close	24,063	43,818	97,598	185,565	280,886	378,946	474,070

Source: Government of India (2010).

What is the administrative procedure for subsidy disbursal, and how is it implemented in practice?

With the intention of ensuring fair and transparent distribution of drip subsidies, the Horticulture Department, the administering agency for the government of Madhya Pradesh, has put in place an elaborate procedure for subsidy disbursal. A general description of the process, along with the time allotted for each step, is provided below. Specific steps may vary in practice; however, the general level of complexity is universal, according to our interviewees.

The process starts with block development officers (blocks are units within Indian administrative districts) receiving a *gram sabah* (village governance unit) approved application and required documents from the farmer through the village horticulture extension officer, horticulture development officer or other sources, such as dealers of registered manufacturing companies. The compiled applications are put in a register, registration numbers and dates are assigned and the beneficiary is notified in writing. Within three days of receiving the application, a visit to the farmer's field is made with the village horticulture extension officer to confirm the information in the application, and field notes are compiled in a folder. Within four days, applications are arranged in an ordered list and sent to the district horticulture office along with an application. A copy of the list and application remain with the block development officer.

The District Micro Irrigation Committee registers the lists by block within three days. The estimated drip area and costs are put into an integrated action plan and sent to the State Micro Irrigation Committee within 10 days. Based on the approvals from the State Micro Irrigation Committees, state-level action plans are prepared and sent within 10 days to the central government's Department of Agriculture & Cooperation in New Delhi. Approvals are then sent back to the District Micro Irrigation Committees within three days.

After receiving the approvals, each registered application is forwarded to the manufacturing company selected by the farmer applicant. The company makes an application to a bank for financing and prepares irrigation designs. The designs are shared with the block development officer and the senior horticulture development officer, with the overall process to be completed within three days.

After receiving the documents, the block development officer visits the farmer's field and makes an initial physical verification of irrigation water and available energy for the proposed system. Once verified, the required documents along with a recommendation are sent to the member secretary of the District Micro Irrigation Committee within three days. The member secretary then has three days to issue administrative approval and gives work orders to the company to begin work. The company has seven days to complete the work, train the farmer in system use and provide a user manual in Hindi. After completing the work, the company has to obtain a certificate of satisfaction from the farmer and send it to the member secretary along with an invoice.

District Micro Irrigation Commission officers must then physically verify the system and confirm that it has been established according to the original plans and uses approved materials in the correct quantities and that irrigation water from the system is uniformly applied. They also confirm that the farmer has been trained and received a user manual. A verification report and approval of payment is written. The entire process is to be completed in 10 days. If the physical verification shows that the requirements have not been met, the member secretary must inform the company and ask for change to bring the system into compliance. This process is allowed three days. The physical verification is then repeated, and if approved, payment is authorized.

Once the work is verified, the farmer recipient must submit an affidavit on a 5-rupee stamp paper, along with required forms, to the assistant director, horticulture, joint member secretary and District Micro Irrigation Committee declaring that s/he will not transfer the system to another farmer or sell it and that s/he and no other combined family members have taken previous benefits from the central or state governments for drip.

In total, the procedure requires farmers to fill out 14–18 forms, attach multiple documents and documentary evidence, obtain 'no-objection' statements and clearances from multiple agencies, and move paperwork between a variety of agencies and government departments on a particular timeline. This interaction with the government must take place concurrent with liaisons with dealers to obtain the required clearances. The extent to which these processes must be adhered to is unclear, though several stakeholders we spoke to suggested large differences between theory and practice, for example with actual processing times much longer than the government guidelines stipulate. Our interviewees reported times of six months to a year.

This process is so complicated and time-intensive as to put even the most enthusiastic drip farmer off the idea of applying for a subsidy. In actual practice, however, there is no need for him or her to do so, as an industry of agents (facilitators) has emerged to move farmer applications through the system. The agents act on behalf of equipment manufacturers or dealers and facilitate the entire subsidy process for farmers, starting with the initial stage of procuring the blank forms, getting them signed by the farmer, submitting the application forms and obtaining necessary approvals at all stages of the process. In fact, all 40 of the drip farmers we interviewed, in multiple districts of Madhya Pradesh, had acquired drip irrigation using an agent. While agents are supposed to assist farmers, the direction of engagement was actually the reverse. Rather than farmers seeking out agents, farmers were sought out by agents, who offered to help them obtain the subsidized drip systems.

Who gains from the current system of subsidy delivery, and what is its impact on manufacturer pricing?

The fact that agents seek farmers rather than farmers seeking agents suggests that there are serious problems with the subsidy system's operation. Farmers indicated that the primary focus of agent visits was to explain the opportunity for the farmers to obtain high-priced products at low cost and to organize the necessary papers for release of the subsidy from the government. The focus was not on the benefits of using the technology or details about maintenance. The agents often promote the lower-quality drip configurations and suppliers within the government-approved lists. This appears to be because dealer margins are greater on lower-quality but still certified products, since all certified products receive the same subsidy. Farmers often accepted the recommendations, perhaps in part due to lack of information, but also in part due to agents' offering faster processing of files and sometimes discounts for the lower-quality choices. Overall, it appeared that the farmer had become a passive participant in the subsidy process, with the agent pushing adoption and making the decisions about product choice and supplier.

If it is true that farmers are less-than-active participants in the subsidy acquisition system, we would also expect to see less-than-enthusiastic use of the products once supplied. Casual observations of the alternative purposes to which drip equipment is put confirmed this. For example, in the villages we visited, we saw drip lines rolled up and hung in trees, turned into webbing for chairs, and made into children's swings. This at least suggests that the primary beneficiaries of drip subsidies may be manufacturers or dealers rather than farmers.

Further evidence that the beneficiaries are not farmers can be found in the prices of subsidy-eligible equipment. To assess the price competitiveness of products sold under the subsidy, one would need to compare the prices of subsidized products with prices of similar products available in the open market. However, such a comparison is constrained by the lack of unsubsidized systems on the market (see below). Nevertheless, we were able to find some comparators for a few components of the drip systems sold under the subsidy scheme. Manufacturers eligible for the subsidy scheme provide an important component of the system, Venturi, at INR 4800–5000 (approximately half the system cost for the smallest farms using wide lateral spacing), while in the open market we were able to find the component for around INR 2400. A filter valued at around INR 11,000 in a manufacturer's subsidy-eligible system was available at around INR 5000 in the open market. Price differences were similar, less than 50% of the 'subsidized' cost, for the few other components we could find. It should be noted that the comparison made here is between wholesale, subsidy-eligible prices and retail, unsubsidized prices, so the apparent discrepancy is probably even greater than these figures indicate. Earlier work by Polak and Sivanapan (1998) also concluded that the subsidy system increased prices, as did the work of Shah and Keller (2002), who estimated that non-approved components were 60–70% less expensive than government-approved components of similar quality.

Dealers and manufacturers we spoke to suggested that the prices of drip systems now on the government price list would fall at least 40% if the current subsidy system were withdrawn by the government and perhaps even further over time as true competition increased between suppliers. Such a drop in prices would not necessarily hurt the higher-quality manufacturers, who report that a large part of their 'gain' from the subsidy system is lost to bribery. For example, an employee of one major manufacturer said that bribes to smooth the approval process make up 30–35% of costs, and that his company offers a discount of this size to customers who buy directly from them rather than working through an agent and the subsidy system.

In addition to working against the goal of reducing costs to encourage investment, the system of subsidy delivery also works against the goal of targeting poor farmers. Disadvantaged farmer groups are eligible for higher subsidies, as discussed above. However, the cost structure built into the current subsidy regime indirectly incentivizes both manufacturers and dealers of the drip system, as well as the government agencies administering the subsidy programme, to target better-off farmer groups. As can be calculated from Table 3, the average cost per hectare for drip systems is higher for smaller plots. In the case of 4 × 4 spacing, for example, the cost per hectare for 0.2 ha farms is more than twice that for 5 ha farms. These cost differentials cannot be blamed on the subsidy system per se. However, by focusing on a smaller number of relatively well-informed and financially better-off farmers for subsidy disbursal, dealers save on advertising and marketing, and the officials administering the subsidies significantly reduce their oversight efforts, including the need for site visits for physical verification, while at the same time having a better chance of meeting their targets for area under drip given their limited budgets. The result appears to be that subsidies preferentially benefit larger farmers, a conclusion also arrived at by Namara, Upadhyay, and Nagar (2005) for Gujarat and Maharashtra and by Palanisami, Raman, and Mohan (2012) for Tamilnadu, Mahrashtra, Rajsthan and Gujarat.

Are alternative approaches possible?

Systematic data on the number of farmers adopting (or disadopting) drip irrigation and the area under drip irrigation are not collected by any agency in India. Currently available official data on both are estimated on the basis of subsidized drip system sales. In essence, subsidized sales are considered the only sales. Independent assessment, based on field visits and discussions in Madhya Pradesh with farmers, agents, manufacturers and district officials, and adopter farmers, confirms that this estimation approach may be reasonable, as it appears that almost all drip equipment sales are subsidy-driven. Manufacturers and dealers, in fact, estimated that more than 95% of the drip sales in Madhya Pradesh are subsidy-linked. Given the nature and magnitude of the subsidy system, this is not unexpected.

Because drip sales occur almost entirely under the subsidy system, one might maintain that those advances which have occurred in drip usage are because capital cost subsidies are necessary. This line of reasoning is advanced not only to justify continuation of the current programme, but also to increase subsidy levels even further and to expand the volume of money in the programme so that more farmers can participate at the higher rates. It is supported by indications from some manufacturers and dealers of unmet demand by farmers, who must now wait one to two years before acquiring a subsidized drip system.

However, if there is unmet demand for drip services, it may be due more to the presence of the current subsidy system than to a lack of subsidy funds. That the subsidy system may hold drip expansion back and that alternative models were needed was already concluded more than 15 years ago by Polak and Sivanapan (1998).

While the subsidy has encouraged some farmers to install drip systems, it has had paradoxical results. For example, delays as long as one year in releasing subsidy payments to manufacturers produce paradoxical price increases for subsidized equipment. Changing the design of effective drip systems so that they can be sold profitably by the private sector at a price lower than the existing subsidized price opens up the possibility of replacing subsidies with an alternative that produces the intended impacts without the disadvantages.

There are many technical, socio-economic and political reasons, beyond capital costs, why drip irrigation has not expanded further in Madhya Pradesh, in India, and world-wide (Namara et al., 2005; Kumar, Turral, Sharma, Amarasinghe, & Singh, 2008; Singh, Patel, Trivedi, & Patel, 2015; Venot, 2016). Even if capital costs are truly a key barrier, there are alternatives to the current subsidy system that would avoid some of its current problems. For example, the government could experiment with more flexible subsidy systems, such as that associated with the Gujarat Green Revolution Company, which requires verification of product purchase and use but gives more flexibility in what is purchased. Alternatively, farmers can either be provided with interest free loans for the entire cost of drip system, administered through existing financial institutions, or provided with conditional cash transfers. Farmers could use the loans or cash to buy drip systems from any dealer or manufacturer, choose the desired configuration of the drip system (e.g. whether or not to include the expensive Venturi), and negotiate price and after-sale service conditions with the dealer. Farmers would not need to visit government offices to obtain approval, clearances or no-objection certificates before buying the system. Agents would therefore no longer have a role, though verification would still be necessary. Since subsidies are now 70–80% of the high approved costs, a price reduction of just 20–30% due to the end of the subsidy system would make the direct government costs of the programme no higher than they are now and eliminate much of the indirect administrative costs.

And even if direct costs did not fall, an end to the current system would bring other benefits. As farmers would be able to determine exactly what system they want, they could choose systems better suited to their own fields, and therefore cheaper or more effective, resulting in indirect cost reductions or profit gains. Choice would also allow farmers to take advantage of new designs, which do not always make it into the subsidy-approved systems. For example, it has been estimated that system costs could be reduced by 25–30% through slight modifications in agro-techniques, such as paired row planting (Palanisami et al., 2012). Finally, the alternative approach directly addresses the initial rationale for drip system subsidies: high initial investment costs. Under the existing system, farmers must still obtain funds to meet their unsubsidized share of the initial (artificially high) costs. Under the alternative scheme, there would be no initial capital requirements for the farmer.

Conclusion

While drip irrigation has been used effectively in many regions of the world, the apparent promise of the technology has frequently met with disappointment for a variety of reasons that go beyond technical promise (Venot, 2016; Venot et al., 2014). The situation is no different in India. Almost half a century since its introduction, drip now covers only 2–5% of its officially recognized potential area. There are a number of possible reasons why a technology with so much promise on paper and in field trials may not take off in practice. One of the commonly cited reasons in India is the high upfront capital costs. In response, the government of India and state governments have made capital-cost subsidies their primary means of drip promotion. The focus on subsidies was reaffirmed after the 2004 National Mission on Micro Irrigation report, and new aggressive adoption targets were set. Despite subsidies of 70% or more provided by the central and state governments, there is little chance the targets will be met.

To understand why the subsidies are not meeting impact expectations, we examined their implementation in Madhya Pradesh, a state with a long history of drip subsidies, high subsidy levels, a large drip potential (1.3 million ha) and low drip implementation (20,000 ha, or 1.48% of potential). We found a number of problems with the subsidy system as currently operated. First, the subsidy regulations force farmers to purchase particular drip system configurations, which may not suit their particular needs, thereby increasing investment costs and reducing benefits. Second, though the systems eligible for subsidy have government certification, that certification is not always granted through standard procedures. Third, the process of obtaining subsidies is so complicated that most farmers are not interested in making the effort on their own, though agents have emerged as middlemen to assist them. Fourth, agents often encourage farmers to purchase lower-quality systems with higher margins for dealers or manufacturers. This is because, fifth, the current system gives manufacturers no incentive to reduce costs and has higher margins for lower-quality but still certified products. Finally, the system favours larger farmers, who are not necessarily the main targets of the programme.

Thus, while the government of India had the good intention of encouraging better water use by subsidizing drip systems for farmers, the result may well be that the subsidy system holds back drip expansion. It has killed incentives to bring down manufacturing costs through alternative product designs or technological innovations. It has brought unfair competition to legitimate drip manufacturers through a flawed certification system. It has forced almost all sales out of the free market and into the bureaucratic system, in effect making manufacturers subservient to government favours and leading to a loss of enterprise spirit, further impeding the growth of a drip market. And it has forced farmers to buy suboptimal systems, discouraging increased use and farmer innovation.

There are arguments that expansion of drip is hindered because subsidy rates and volumes are insufficient. Given the problems identified here, it is equally possible that drip would expand more quickly if the current subsidy system were removed. If capital costs truly are an issue, alternative approaches are possible. The Gujarat Green Revolution Corporation gives farmers more flexibility and is perceived by farmers to reduce corruption (Bhamoriya & Mathew, 2014). An interest subsidy programme that placed loans in the hands of farmers rather than subsidized products in the hands of dealers might also address some of the current programme's worst shortfalls. While there would be a net cost to farmers if the drip market did not change in response, we believe that substantial price reductions would occur, offsetting a large part of the increased costs. The increase could be completely offset if the government transferred its gains from the new programme back to farmer recipients. Government costs would not increase; farmers would have more choice; and the process would be more transparent, less prone to corruption and easier to manage and govern. While this study was limited to the state of Madhya Pradesh, the findings probably apply widely in India.

Disclosure statement

No potential conflict of interest was reported by the authors.

References

Bhamoriya, V., & Mathew, S. (2014). *An analysis of resource conservation technology: A case of micro-irrigation system (drip irrigation)*. Ahmedabad: Centre for Management in Agriculture, Indian Institute of Management. August 2014.

Government of India. (2004). *Salient findings and recommendations of task force on micro irrigation*. New Dehli: Ministry of Agriculture, Department of agriculture and Cooperation.

Government of India. (2010). *National mission on micro irrigation: Operational guidelines*. New Delhi: Ministry of Agriculture, November 2010.

INCID. (1994). *Drip irrigation in India*. New Delhi: Indian National Committee on Irrigation and Drainage.

International Commission on Irrigation and Drainage [ICID]. (2012). Sprinkler and micro irrigated area. Retrieved June 12, 2016, from www.icid.org/sprin_micro_11.pdf

Kumar, M. D., Turral, H., Sharma, B., Amarasinghe, U., & Singh, O. P. (2008). Water saving and yield enhancing micro irrigation technologies in India: When and where can they become best bet technologies? 7th Annual Partners' Meet of IWMI-Tata Water Policy Research program, ICRISAT, Hyderabad.

Namara, R., Upadhyay, B., & Nagar, R. K. (2005). *Adoption and impacts of micro irrigation technologies: Empirical results from selected localities of Maharashtra and Gujarat states of India* (Vol. 93). Colombo, Sri Lanka: IWMI.

Narayanamoorthy, A. (2009). Drip and sprinkler irrigation in India: Benefits, potential and future directions. *India's water future: Scenarios and issues. Strategic Analyses of National River Linking Project of India. Series, 2*, 253–266.

Narayanamoorthy, A. (2012). Drip method of irrigation in Maharashtra: Status, economics and outreach. In K. Palanisami, S. Raman, & K. Mohan (Eds.), *Micro-irrigation: Economics and outreach* (pp. 120–139). India. New Delhi: MacMillan Publishers.

Palanisami, K., Mohan, K., Kakumanu, K. R., & Raman, S. (2011). Spread and economics of micro-irrigation in India: Evidence from nine States. *Economic and Political Weekly, 46*, 81–86.

Palanisami, K., Raman, S., & Mohan, K. (2012). *Micro irrigation- economics and outreach*. New Delhi: MacMillan Publishers India. ISBN 978-935059-062-1.

Polak, P., & Sivanapan, R. K. (1998). The potential contribution of low cost drip irrigation to the improvement of irrigation productivity in India. *India – Water Resources Management Sector Review, Report on the Irrigation Sector*. The World Bank in Cooperation with the Ministry of Water Resources, Government of India. (Vol. 28, p. 30).

Raman, S. (2010). *State-wise micro irrigation potential in India- An assessment*. (un published). Natural Resources Management Institute, Mumbai.

Shah, T., & Keller, J. (2002). Micro-irrigation and the poor: A marketing challenge in smallholder irrigation development. In H. Sally & C. L. Abernethy (Eds.), *Private irrigation in sub-Saharan Africa: Regional seminar on private sector participation and irrigation expansion in sub-Saharan Africa, 22-26 October 2001, Accra, Ghana* (pp. 165–184). Colombo, Sri Lanka: International Water Management Institute, Food and Agriculture Organization of the United Nations, and ACP-EU Technical Centre for Agricultural and Rural Cooperation.

Singh, P. K., Patel, S. K., Trivedi, M. M., & Patel, G. R. (2015). Assessing the relative impacts of the factors affecting MIS adoption process. *International Journal of Sustainable Development & World Ecology, 22*, 213–218.

van der Kooij, S., Zwarteveen, M., Boesveld, H., & Kuper, M. (2013). The efficiency of drip irrigation unpacked. *Agricultural Water Management, 123*, 103–110.

Venot, J. P. (2016). A success of some sort: Social enterprises and drip irrigation in the developing world. *World Development, 79*, 69–81.

Venot, J. P., Zwarteveen, M., Kuper, M., Boesveld, H., Bossenbroek, L., Kooij, S. V. D., & Verma, S. (2014). Beyond the promises of technology: A review of the discourses and actors who make drip irrigation. *Irrigation and Drainage, 63*, 186–194.

Managing water-related risks in the West Bengal Sundarbans: policy alternatives and institutions

Ernesto Sánchez-Triana, Leonard Ortolano and Tapas Paul

ABSTRACT

Persistent pressures from water-related threats – sea-level rise, soil and water salinization, and flooding due to embankment overtopping and failure – have made the West Bengal Sundarbans a challenging place to live, and effects of global climate change will only worsen conditions. Four alternative policy directions are examined: business as usual; intensive rural development; short-term out-migration of residents; and embankment realignment and facilitation of voluntary, permanent out-migration. The last of these is the recommended approach. Study findings have informed ongoing deliberations to build consensus on future policy directions for reducing the region's vulnerability to natural disasters.

Introduction

The Sundarbans region (Figure 1), which is formed by the deltas of the Ganges, Brahmaputra and Meghna Rivers, contains what is arguably the world's largest remaining mangrove. The forest is home to many endangered species, including the iconic royal Bengal tiger. About 40% of the region lies within India, and the remainder is in Bangladesh. It is bounded on the west by the Bhagirathi-Hooghly River, a tributary of the Ganges, and on the east by the Padma-Meghna River, which flows from the Brahmaputra. Although the Sundarbans lies in both India and Bangladesh, this article treats only the Indian Sundarbans, which is entirely within the state of West Bengal. (Unless otherwise noted, 'Sundarbans' refers to the Indian Sundarbans.)

The total area of the Indian Sundarbans is 9630 km^2, of which 4264 km^2 consists of wetlands and mangrove (Hazra, Ghosh, DasGupta, & Sen, 2002). While the area was once entirely covered by mangrove, much of it has been converted to other uses, such as agriculture and aquaculture. Of the hundred or so islands that are currently part of the Indian Sundarbans, 54 are inhabited to varying degrees, and the others are forested (Danda, 2007). The population of the Sundarbans rose from nearly 1.2 million in 1951 to almost 3.2 million in 1991 and to 4.4 million people in 2011 (Hazra et al., 2002; Sánchez-Triana, Paul, Ortolano, & Ruitenbeek, 2014).

The supplemental material for this article is available online at http://dx.doi.org/10.1080/07900627.2016.1202099.

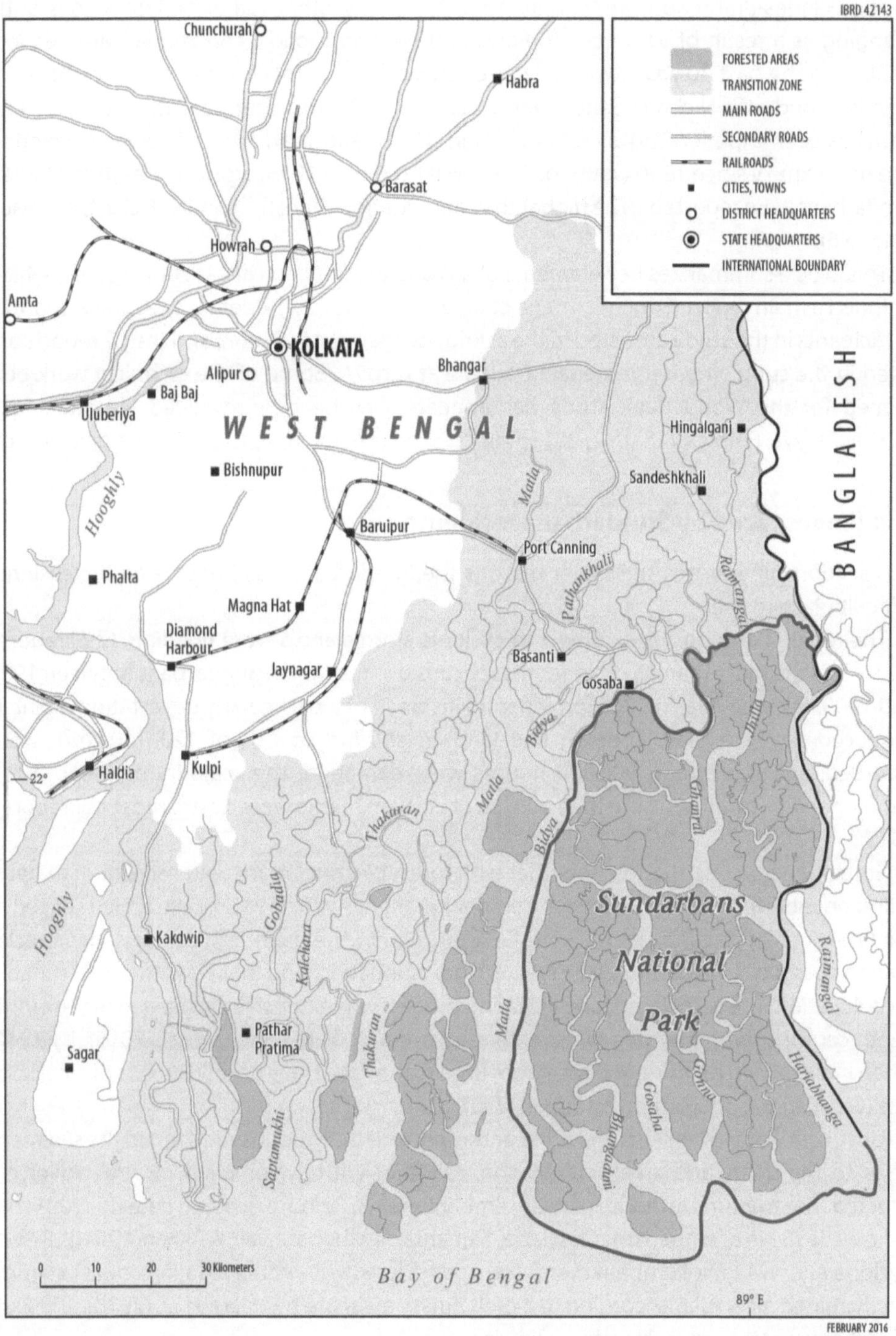

Figure 1. The Sundarbans region of India. Figure prepared by the Map Design Unit of the World Bank (see online version for colours).

The Ganges-Brahmaputra-Meghna Delta, like many other major deltas, is constantly changing as a result of variations in flows and sediment loads. The Sundarbans has lost 100 km^2 in the past 30 years, with effects of erosion felt mainly on the southern shores of islands (Woodroffe, Nicholls, Saito, Chen, & Goodbred, 2006). In contrast, the Meghna Delta plain has been experiencing accretion. Allison (1998) reports an average net land accretion rate of 4.4 km^2/y since 1840 (south of 22.90 °N latitude). In 2015, erosion rates in the Indian Sundarbans were reported to be higher than aggradation (Ghosh, Schmidt, Fickert, & Nüsser, 2015, § 6).

This article summarizes key elements of a World Bank technical-assistance study, which included a main report (Sánchez-Triana et al., 2014) as well as 12 detailed technical reports. Participants in the study are listed in the acknowledgements section, and the 12 reports are listed in the supplemental material (available at http://). Some of the technical work performed for the World Bank study has appeared in the peer-reviewed literature (e.g. Bhattacharyya, Pethick, & Sarma, 2013; Biller & Sanchez-Triana, 2013), but most of it has not.

Challenges faced by Sundarbans residents

This section summarizes the major reasons the Sundarbans has become an increasingly difficult place to live.

The region has long been subject to cyclonic storms and severe flooding. Chakraborty (2015) lists the more than 35 cyclones that occurred in the Indian Sundarbans between 1909 and 2009. Cyclone Aila, in 2009, provides an instance of a particularly devastating event. It produced heavy rainfall, flooding and landslides. Wind speeds of 120–140 km/h were recorded, and more than 900,000 houses were damaged, the majority of them in the Sundarbans. In addition, more than 500 km of embankments were destroyed (Chakraborty, 2015).

Embankment construction, together with forest clearing, made extensive human habitation possible in the Sundarbans. The embankment construction process, which started in the late nineteenth century and continued through to the twentieth century, made reclamation possible by preventing saline water from inundating land that was otherwise suitable for cultivation. The associated forest clearing process was extensive. At the end of the eighteenth century, the mangrove forest extended up to Kolkata (Ghosh et al., 2015); by 1947, when India gained its independence from Britain, the forest was only 50% of its pre-colonial size (Giri, Pengra, Zhu, Singh, & Tieszen, 2007).

Upstream water diversions and other anthropogenic activities have changed the sediment flows to the delta, and creekside erosion, together with cyclonic storms, has adversely affected the structural integrity of many embankments. Remote-sensing data (for 2001–08) indicate 3638 km of embankments (Hazra, Samanta, Mukhopadhyay, & Akhand, 2010). Before Cyclone Aila, 471 km of embankments were considered vulnerable to failure, but that number increased after Aila. According to a preliminary estimate by Hazra et al. (2010), 1000 km of embankments were vulnerable following Aila.

In addition to the effects of creekside erosion in undermining embankments, many of those living behind embankments face frequent challenges associated with embankment overtopping due to storm surges caused by cyclones. For example, in 2005, the western side of Mousuni Island lost about 1.5 km of embankment to storm surges (Danda, 2007). A study of 600 of the 1550 families in one village (Baliara) on Mousuni Island found that nearly 13%

of respondents had been adversely impacted by storm surges on an annual basis (Danda, 2010). Such surges typically take place when very high winds occur simultaneously with spring tides. It is difficult to drain the relatively low land behind the embankments after they are breached or overtopped, and this has led to increased soil salinity, with consequent challenges in farming the land (Ghosh et al., 2015; Raha et al., 2013).

Sea level, which has been rising for millennia in the Sundarbans, is predicted to make matters worse. Much recent attention has been given to sea-level rise (SLR), especially global (sometimes called eustatic) SLR. Eustatic SLR refers to changes in the mass of the oceans due to the melting of glaciers and other land-supported ice, as well as the thermal expansion of seawater as the oceans warm (Ericson, Vörösmarty, Dingman, Ward, & Meybeck, 2006). Changes in salinity also play a small role in global SLR (Antonov, Levitus, & Boyer, 2002).

However, eustatic SLR is only one component of relative SLR, i.e. the rate of apparent sea-level change relative to the land surface. In addition to eustatic SLR, relative SLR in the Indian Sundarbans is influenced primarily by land subsidence, e.g. from auto-compaction, tectonic activity and anthropogenic processes, including groundwater withdrawals (Bhattacharyya et al., 2013). Further information on land subsidence contributing to SLR, as well as challenges resulting from the limited relevant data in the Sundarbans, is given in the supplemental material.

As detailed later in this article, the embankment system that was intended to protect the Sundarbans from the sea is not working well, and many kilometres of embankments are vulnerable to breaches and overtopping due to storm surges. Based on extensive modelling studies, Kay et al. (2015) predict that flooding due to storm surges during increasingly high tides will be an even more significant risk for residents of the Ganges-Brahmaputra-Meghna Delta, which includes the Sundarbans.

Changes in salinity in the estuaries of the Indian Sundarbans are another concern. Owing to higher freshwater flows in the Hooghly Estuary, salinity has decreased in that zone. As a consequence, fish species of low commercial value, particularly hilsha (*Tenualosa ilisha*), have increased significantly, but the commercially important taxa have become less abundant (Ghosh, 2015; Sinha, Mukhopadhyay, & Mitra, 1997). The situation is different in the central portion of the Sundarbans. Salinity has increased in that area because connections to melt-water sources have been eliminated by heavy siltation of the Bidyadhari Channel (Banerjee, 2013).

Although net forest area has not diminished much in recent decades, the changes in the mix of tree species have been notable (Giri et al., 2007). Part of this shift in species is explained by forest replanting programmes and by shifts to salt-tolerant mangrove in areas of rising soil and water salinity (Giri et al., 2007; Hazra et al., 2010). Also, increases in pollution and soil salinization have caused top-dying disease of *Heritiera fomes* (commonly known as sundri), the once-dominant and economically most important mangrove species in the Sundarbans (Ghosh et al., 2015). Of special concern are the central portions of the forest, where levels of salinity have increased (Banerjee, Roy Chowdhury, Sengupta, Sett, & Mitra, 2012a). In addition, remote-sensing studies for 2001–08 show continued land conversion, with agricultural land being converted to dense settlement areas and aquaculture ponds; the area of agricultural land dropped in that period from 2149 km^2 to 1691 km^2 (Hazra et al., 2010).

The drop in land suitable for cultivation has been accompanied by an increase in shrimp farming, and many residents are supplying tiger shrimp seeds for aquaculture. This was the case for 20% of the 243 households surveyed by Danda (2007). However, seed collection practices result in significant aquatic ecosystem disruptions as a result of bycatch. The target species (*P. monodon*) accounts for at most 1% of the total catch brought from the sea during seed collection, and almost of all the remaining catch is discarded on beaches or tidal mudflats (Danda, 2007). Capture of aquatic species caught unintentionally during shrimp seed collection, together with overexploitation of shrimp fry, is severely limiting the numbers of fully grown shrimp in natural waters (Knowler et al., 2009).

Pollution from Kolkata has increased, as evidenced by heavy metal contamination of Hooghly Estuary. For example, Samanta et al. (2005) found Cd, Cu and Pb at levels high enough to disturb aquatic life processes in the portions of the Hooghly near Haldia. In addition, Guzzella et al. (2005), found polychlorinated biphenyls and other persistent toxic pollutants in sediments along the lower stretch of Hooghly Estuary. Much of this pollution has been linked to sources in Haldia and Kolkata (Banerjee, Senthilkumar, Purvaja, & Ramesh, 2012b; Samanta et al., 2005).

Other challenges in the Sundarbans concern increases in human–animal conflicts. In the case of tigers, this is because of habitat loss and the encroachment of humans into tiger territories. Das (2014) conducted household surveys in villages adjacent to the mangrove forest and identified 237 incidents of tigers straying into those villages during 1995–2010. In the context of that survey, the vast majority of incidents involved predation on livestock, followed by injury to people. There were also 7 cases of human death and 12 cases in which villagers killed tigers.

Rationale for the proposed strategy

This section summarizes the approach used to develop a strategy for reducing the risks faced by Sundarbans residents and conserving biodiversity, and subsequent sections of the article highlight elements of the strategy.

The World Bank study that provides the basis for this article was conducted from 2009 to 2012. The study team included World Bank staff and consultants, who prepared 12 different background papers across a broad range of subjects: livelihoods dependent on aquaculture, agriculture and ecotourism; education; health; water and sanitation; energy; transportation; and household air pollution from use of solid fuels. Also, a survey of 2188 households in the Sundarbans was conducted to provide a demographic and economic profile of households and information on patterns of migration in response to extreme-weather events (Ortolano, Sánchez-Triana, Paul, & Ferdausi, 2016). A study dealing with weather-related hazards identified cyclonic-storm trends. A study on coastal geomorphology looked at changes in erosion and accretion in tidal channels and consequent effects on the integrity of embankments. Furthermore, studies were conducted of the effectiveness of existing disaster risk management systems and government subsidies for Sundarbans residents (Sánchez-Triana et al., 2014).

Literature reviews informed all aspects of the study, although several reports involved primary data gathering. In the context of this article, primary data gathering was done by Bhattacharyya et al. (2013) for the geomorphological modelling studies summarized herein. In addition, the household survey, which was conducted by Economic Information Technology, with participation by Professor Guatam Gupta of Jadavpur University in Kolkata, also involved primary data gathering. The survey results were detailed in an unpublished study, and key aspects of the results (as well as information on the household sampling strategy) were summarized by Ortolano et al. (2016).

In addition, workshops and consultations were held with representatives of multiple levels of government. At the national level, this included India's National Disaster Management Authority, Ministry of Surface Transport in Kolkata, Ministry of Shipping, Inland Waterways Authority and National Highway Authority. At the West Bengal level, those consulted included the Department of Planning, Sundarbans Development Board (SDB) and members of the West Bengal Legislative Assembly representing local constituencies. Other consultations were with local research institutions (e.g. Department of Oceanography, Jadavpur University); nongovernmental organizations (NGOs) such as the World Wildlife Fund; and individual residents. Interactions with stakeholders were used to identify issues to be addressed in individual studies and to discuss findings from the studies as they emerged. These interactions also informed the creation of the proposed strategy.

The study team was motivated to develop a strategy for risk reduction and biodiversity conservation for the following reason. The Indian Sundarbans has become increasingly hazardous, and high levels of material poverty have been evident for generations. For centuries, large parts of the forest have been extensively exploited for timber, fish, and shrimp seeds or converted for agriculture and aquaculture. Upstream water diversions and other anthropogenic activities have changed the landforms in the delta and the structural integrity of embankments. The region's population growth has led to degradation of the mangrove forest, unsustainable extraction of natural resources, and an increase in the number of people exposed to significant and recurring flood risk. Pollution from Kolkata has increased. In addition, the absence of adequate physical infrastructure in the region contributes to low standards of living. Increased salinity and waterlogging has imperilled agriculture, and shrimp farming practices have been ecologically unsustainable. Moreover, the frequency of human–animal conflicts has increased because of habitat loss.

The portion of the Sundarbans in which residents face particularly high risks and hardships, referred to herein as the 'transition zone', is between the part of the Sundarbans on the mainland (the 'stable zone') and the 'core zone'. The latter is defined herein as the contiguous legally protected areas of the Sundarbans, which includes the Sundarbans National Park and Tiger Reserve, one of the three wildlife sanctuaries in the region, and the Sundarbans Reserve Forest (SRF). The three zones are shown in Figure 2.

The people in the transition zone suffer from a shortage of livelihood opportunities. They also lack the economic, human and social capital to make permanent out-migration to urban job centres successful in the short term. Job training would be a necessary precondition for successful out-migration.

The study team considered four approaches to improve the condition of transition-zone residents: business as usual; intensive rural development; short-term out-migration of residents; and embankment realignment and facilitation of voluntary, permanent out-migration. The business-as-usual approach is a continuation of the current low-intensity rural development strategy, with influxes of aid following disasters. It does nothing to effectively address the region's challenges and will only perpetuate the ongoing cycle of poverty. A more intensive rural development approach was considered and rejected because it would attract migrants from outside the Sundarbans and thereby put more people at risk; it would also put additional pressure on the mangrove forest. The third of the approaches examined involved short-term migration of residents out of the transition zone. This was ruled out

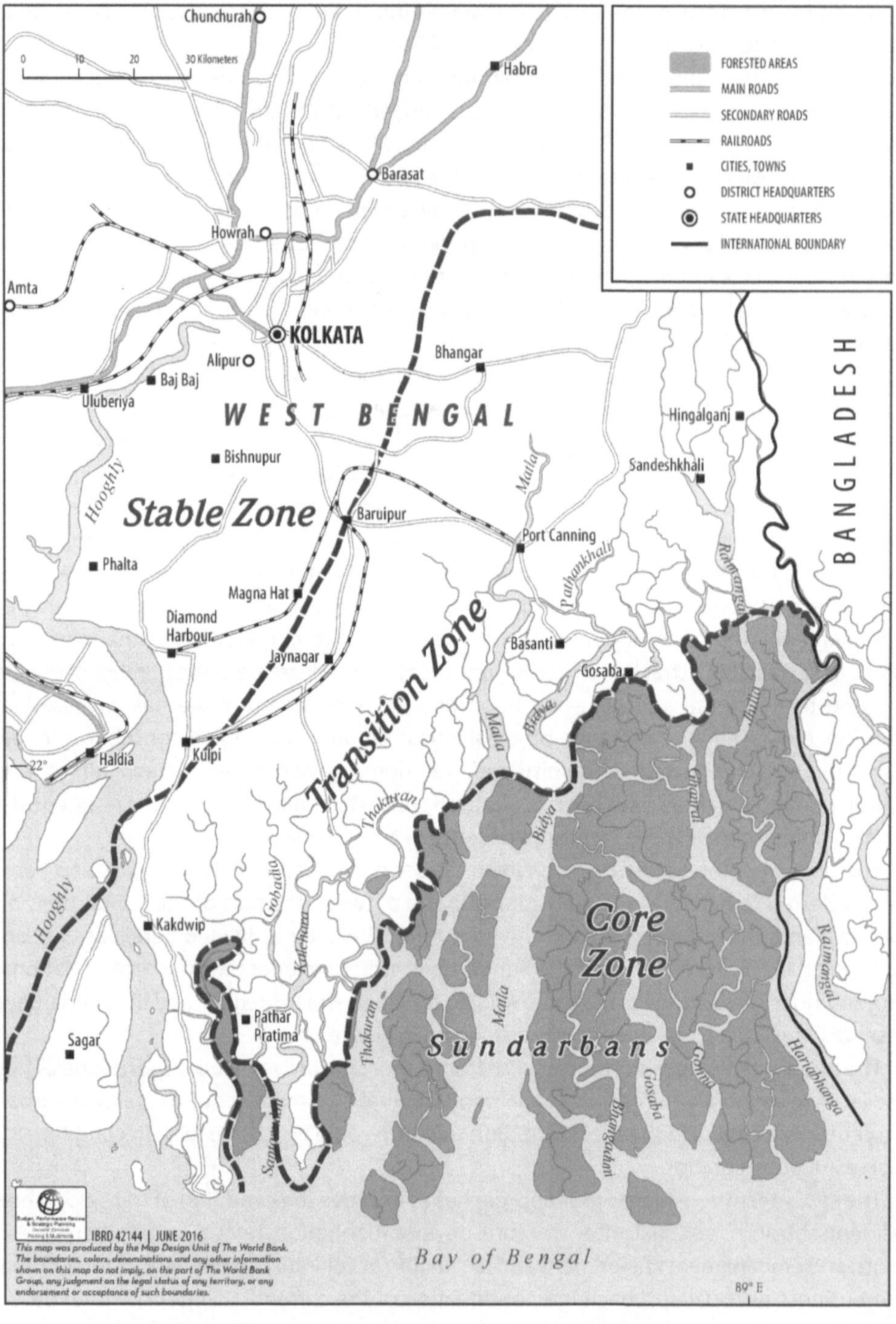

Figure 2. Map of the West Bengal Sundarbans, showing the core, transition and stable zones. Figure prepared by the Map Design Unit of the World Bank (see online version for colours).

because there would not be sufficient time for residents to obtain the training and other resources needed to migrate successfully, and the approach would entail massive social disruption.

Based on its analysis, the team recommends the fourth strategy, which reduces both human and ecosystem risks by encouraging the gradual, voluntary movement of transition-zone residents to the stable zone and urban areas outside the Sundarbans that offer more safety and better employment opportunities. This strategy recognizes that transition-zone residents will continue to face stresses, such as embankment overtopping, and that some existing livelihood options are unsustainable. In addition, major investments in transport networks and electricity in the transition zone would be economically inefficient because projected SLR is likely to make parts of the transition zone uninhabitable in coming decades.

The proposed strategy rests on measures to build the economic, human and social capital residents require to migrate permanently outside the transition zone into relatively safe urban areas in the stable zone and beyond. With appropriate planning in urbanized areas, there would be improved options for health care, education and employment. Transition-zone residents have generally not pursued permanent out-migration as an option, in part because they lack the economic, human and social capital required for successful out-migration. To buy time while the needed capital is being created (e.g. using job training programmes) and to reduce risks for those who choose not to migrate away from high-risk areas, the study team proposed measures to improve the integrity of the embankment system and the effectiveness of the disaster risk management system; it also suggested actions to improve the sustainability of livelihood options for those who choose to remain.

At first glance, it may appear internally contradictory to encourage out-migration from the Sundarbans and at the same time reduce water-related risks in the Sundarbans. However, the approach is internally consistent once time scales are taken into account. It will take time to equip transition-zone residents who choose to migrate with the economic, human and social capital needed to take advantage of the greater employment and other opportunities in lower-risk urbanized areas. While this capacity-building process is going on, residents will continue to be exposed to considerable risk, and thus risk-reduction measures are appropriate: realigning embankments gradually and systematically; allowing mangrove to occupy land vacated by realignment of embankments; and improving emergency management in response to cyclones and floods (e.g. improving disaster warning systems and building additional shelters).

The remainder of this article emphasizes the dimensions of the proposed approach most closely tied to reduction of hazards in the transition zone caused by SLR and cyclonic storms. Four aspects of the proposed approach were developed by the team:

- *Modify embankment system and reduce disaster risk*. It is infeasible to eliminate exposure to hazards for transition-zone residents, but risks can be reduced by improving disaster risk management programmes and modifying the embankment system.
- *Enhance the capacity of residents to migrate out of the transition zone*. Steps can be taken to improve the ability of transition-zone residents to migrate successfully, and incentives can be provided for out-migration.
- *Conserve biodiversity*. Biological conservation measures can provide opportunities to reduce *both* hazard exposure (via ecosystem service provision) and vulnerability (by

creating new revenue streams that support building the capacity of transition-zone residents to migrate successfully).

- *Strengthen institutions*. Improvements can be made in the ability of government agencies to implement elements of the proposed strategy.

All aspects of the proposed approach are considered below, but the emphasis is on reduction of hazards in the transition zone caused by SLR and cyclonic storms.

Reducing exposure to climate-related hazards

The proposed approach employs two main types of interventions to deal with the flood-related disasters in the transition zone: short-term measures that extend India's disaster management activities; and a long-term programme of large-scale embankment realignment.

Disaster preparedness

The aforementioned survey of 2188 households illustrates particular challenges concerning flood warnings. Among the households affected by a major cyclone or flood in 2006–11, only 3.6% were notified of the event beforehand (Ortolano et al., 2016). This was the case even though the extreme-weather events in question caused 55.4% of surveyed households to evacuate their homes. Of the residents who faced evacuation, only 2.2% went to shelters (and 26.0% relocated to schools). Moreover, 42.8% of those who were evacuated went to the nearest embankment or elevated road, though these are among the most exposed places during cyclonic storms.

The World Bank study also examined the adequacy of the existing shelter system. It assumed that one cyclone shelter serves 2000 residents, which is typical for the region. Thus, to serve the more than 4.4 million people in the Sundarbans, at least 2000 shelters would be needed. However, as of 2010, only 315 shelters were located in the coastal areas of North and South 24 Parganas (which includes the Sundarbans) and Midnapore East (Basu, 2010). The study recommended a substantial increase in the number of shelters, as well as improved cyclone/flood warning systems and increased interagency and agency–NGO communication during cyclone and flood disasters.

Improvements are being made. The West Bengal Disaster Management Department and Sundarbans Affairs Department are putting in place cyclone shelters, cyclone-resistant buildings, and communication networks to improve response to extreme-weather events. In 2015, West Bengal took actions consistent with those suggested by the study team by joining the National Cyclone Risk Mitigation Project (partially funded by the World Bank). This will provide the government of West Bengal resources needed to build more shelters and enhance its capacity to respond to disasters (National Disaster Management Authority, 2015).

Embankment system challenges

A major effort to reduce hazard exposure in the Indian Sundarbans involves flood-protection measures using embankments, but the embankment approaches currently in use are not working as planned. The vulnerability of the Sundarbans to damaging surges linked to effects of cyclones and SLR is made clear by events associated with Cyclone Aila in May 2009. Storm surges of 2–3 m swept through the region; more than 400 km of embankments were breached; and the entire Sundarbans biosphere reserve was inundated with 2–6 m of water for several days. Damages were estimated at USD 550 million, and 2.5 million people were affected (Chatterjee et al., 2013).

Similar outcomes were observed in the Bangladesh Sundarbans. For example, as reported by Auerbach et al. (2015), during Cyclone Aila, there were several breaches in the embankment at Polder 32. The breaches, which occurred in several places, apparently resulted from undercutting of embankment foundations by erosion. As a consequence of embankment construction, land elevation inside Polder 32 was more than a metre below the elevation of adjacent land outside the polder, and this elevation difference contributed to damage resulting from the embankment breaches.

Based on their assessment of data for the Indian Sundarbans, Bhattacharyya et al. (2013) estimate that considering the *combined* effects of subsidence and eustatic SLR, relative sea level in the Sundarbans is rising at between 3 and 8 mm/y, possibly even more, depending on the location. Higher rates have been observed at the landward edge of the delta (the Kolkata tide gauge), and lower rates at the seaward edge (Sagar Island gauge). Bhattacharyya et al. reported the period of record for both gauges as 1932–1999.

Many others have examined the effects of SLR in the region, and results generally point to significant challenges in the future. For example, modelling studies by Kay et al. (2015) estimated that sea level could rise in the Ganges-Brahmaputra-Meghna Delta by 0.63–0.88 m by 2090, with predictions by others being 0.5 m higher if potential melting of the West Antarctic ice sheet is included. The modelling results led to the conclusion that "climate change could lead to large areas of land being subject to increased flooding, salinization and ultimate abandonment in West Bengal, India, and Bangladesh" (Kay et al., 2015, p. 1311).

Based on field visits after Cyclone Aila, Bhattacharyya et al. (2013) observed that overtopping and wave erosion were not responsible for most breaches in the Indian Sundarbans. Instead, the breaches were caused by mass failures that occurred because of (1) over-steepening of embankment faces as a result of long-term channel erosion; and (2) seepage during surge events, which led to increased pore pressure in the clay structure. These researchers found that channel erosion undercutting of embankment intertidal footings plays a key role in failures because it steepens slope angles, increasing embankment instability during storm surges when pore pressures in the embankments are high.

One component of the World Bank study consisted of modelling conducted by Bhattacharyya et al. (2013). This work employed a regime-modelling approach developed for use in estuaries, as detailed by Wallingford, ABPmer, and Pethick (2006). In this context, 'regime' refers to the *equilibrium* morphology of an estuary channel in response to a tidal discharge. An estuarine channel that neither erodes nor deposits sediments, over a long period of time, is said to be in equilibrium. The central idea in the modelling study was to determine whether particular estuaries in the Sundarbans were wider or narrower than their equilibrium values so that judgements could be made about the suitability of existing embankment locations. Channels narrower than their equilibrium widths could be expected to erode as they approached equilibrium; in contrast, channels wider than equilibrium widths could be expected to accommodate the deposition of sediments. The modelling effort is summarized in the supplemental material.

The key prediction from the modelling effort is the theoretical, time-independent, equilibrium value of channel width; not all estuarine systems under study were expected to be in equilibrium, as they might be either be eroding or accreting. Creeks that are eroding are of particular concern since embankments are located at or near creek edges. Bhattacharyya et al. (2013) used model results to categorize the estuaries based on whether they were narrower or wider than their equilibrium values under prevailing conditions; model runs were also performed for hypothesized future increases of 1 and 2 m in sea level. The estuaries under study were categorized as:

- *Oversized*. These estuaries could accommodate future SLR because reclamation of intertidal lands, together with removal of headwater creeks from tidal flow (using flap sluices), had cut tidal prism volumes.
- *Equilibrium*. Channel widths for these estuaries have kept pace with past SLR, but they contain minimal space to accommodate future increases in sea level, and thus collapse of existing embankments along these channels is expected.
- *Undersized*. These estuaries are undersized as a result of previous SLR and recent increases in tidal flow to and from aquaculture ponds. The recent rapid increase in aquaculture acts to reverse the process of reclamation. Given that accommodation space in these estuaries is already negative, response to future SLR is expected to be severe.

The above-mentioned modelling results suggest that retreating many embankments away from tidal channels can prevent channel bank erosion from undercutting their foundations; this embankment realignment strategy was also highlighted in related studies in the Bangladesh portion of the Sundarbans (Pethick & Orford, 2013). Results from these modelling studies, as elaborated in unpublished technical reports by Pethick (a consultant on the World Bank study) and summarized by Sánchez-Triana et al. (2014), were used in making specific recommendations to the government of India regarding the need for systematic retreat of selected embankments. In some cases, the recommendation was to accommodate future SLR by moving embankments back by as much as 100–300 m (or 50–150 m from each bank of a channel) over the next 20 years and raising the height of a number of embankments.

The managed realignment of embankments near the edges of eroding channels would require that existing embankments be maintained while new ones are constructed further from those channels. In relocating embankments in danger of being undermined by erosion, the land between old embankments (eventually to be abandoned) and new (retreated) embankments will become unprotected. As noted by Bhattacharyya et al. (2013), using an additional setback distance of 30–50 m would allow mangrove to regenerate in these areas, providing a buffer to dissipate the energy of tidal currents and storm-driven winds.

Incentivizing out-migration

SLR, salinization of soil and water, cyclonic storms and flooding have combined over the past century to render the Sundarbans one of the most hazardous parts of the Indian subcontinent. Natural stresses have been compounded recently by human-induced stresses, which include reductions in freshwater flows to many parts of the delta and expansion of tidal-water aquaculture. The predicted changes associated with global climate change will exacerbate these problems: intense cyclonic storms coupled with continuing SLR will further increase exposure to hazards of the local population.

In the long term, trying to address these challenges by following a business-as-usual development scenario will make matters worse, because huge numbers of Sundarbans' residents will remain in harm's way. Moreover, increasing residents' income or building extensive infrastructure throughout the region will only attract more people to areas that are fundamentally in decline and hostile to human habitation. In the long term, voluntary, permanent out-migration from the transition zone appears to be the most viable option.

The suggested measures for improving the disaster response system and realigning embankments will reduce risk in the short and medium term, but these measures will not eliminate risk. Given the risks, increasing numbers of transition-zone residents are likely to view permanent out-migration from the area as a desirable option. This is reflected in increased out-migration during the last few years from the inhabited parts of the Sundarbans closest to the sea, e.g. parts of Sagar Island (Eaton, 2015). The prospect of increased out-migration to cities outside the transition zone is beginning to be discussed by West Bengal officials (Ghosh, 2015).

In recent years, much of the out-migration from the Sundarbans has been temporary. A portion of the aforementioned 2011 household survey probed the migration issue and found that that 31.7% of residents in surveyed households moved in search of work after Cyclone Aila, but that more than 98% of those who migrated did so on a temporary basis (Ortolano et al., 2016). These findings are consistent with the findings of Penning-Rowsell, Sultana, and Thompson (2013) from focus groups in rural Bangladesh: unless a place becomes permanently uninhabitable, people may migrate temporarily in response to cyclones and other natural disasters, but they frequently return to their villages.

While involuntary out-migration could be achieved within a few years, the approach recommended by our study would be socially less disruptive, as it would encourage the gradual, voluntary and permanent out-migration of residents living close to the SRF boundary and other high-risk places to cities in the stable portions of the Sundarbans and beyond that are relatively safe and have greater economic opportunities. Encouraging residents living near the SRF to seek opportunities elsewhere will also reduce the number of residents who illegally exploit the forest's resources.

Most transition-zone residents lack the education and social capital to obtain attractive employment elsewhere, and many are aware of the challenges of life in urban slums. Under the circumstances, what would encourage those residents to migrate? Certainly, providing additional information to residents about the increasingly hazardous conditions and the unsustainability of shrimp farming and other current livelihood options would provide some incentive to leave. Out-migration could also be incentivized by re-evaluating the existing post-disaster government and NGO aid programmes, which encourage residents to rebuild their dwellings and remain in hazardous areas.

Other incentives to leave the transition zone relate to the attractiveness of opportunities in cities for better employment, healthcare and education. Projections by McKinsey Global Institute (2010) show that the bulk of India's growth in both population and GDP during the coming decades will occur in urban agglomerations. We propose the use of small subsidies to reduce the risks associated with moving to an unknown place on a journey with uncertain outcomes. The effectiveness of small subsidies to encourage out-migration in rural Bangladesh was illustrated by Bryan, Chowdhury, and Mobarak (2014).

Incentives to migrate will not be enough. There also needs to be a programme of education and skills development to enhance the likelihood of success of migrants to new, more urbanized locations. In many such locations, there are opportunities for plumbers, electricians, mechanics, skilled masons, etc. Improved education and vocational training is a prerequisite for labour-market success.

Given that Sundarbans residents in the transition and stable zones will need training to fill jobs offered in livelihood clusters and more densely populated parts of the stable zone as well as in urban areas outside the Sundarbans, education and skill-building programmes could be created (in the stable zone) to train the labour force for new manufacturing and service-sector jobs (e.g. hotel workers). These can be targeted to transition-area residents who would be willing to relocate to seek better jobs.

Notwithstanding the challenges facing migrants in Kolkata and other large cities, those areas offer better opportunities for employment and education compared to the transition zone. Moreover, over multiple generations, descendants of the rural poor who seek employment in urban areas are often better off than they would have been if their families had remained as rural residents (Glaeser, 2011; Narayan, Pritchett, & Kapoor, 2009; Ravallion, 2007). In addition, urban economic growth can positively impact the living standards of those who remain in rural areas not only because of remittances sent by family members in cities, but also because fewer people are left in those areas to compete for already scarce jobs.

Out-migration can also be incentivized using conditional cash transfers in the form of health and education vouchers usable in cities in India outside the transition zone. Some results of the aforementioned 2011 household survey suggest that this approach could be successful: many respondents expressed willingness to send their children outside the Sundarbans for educational opportunities (86%) or to participate in training programmes in Kolkata or other parts of West Bengal (78%).

The government of West Bengal and local governments could prepare for the anticipated increase in rural-to-urban migration by promoting rational urban land use and building needed infrastructure (particularly in transport and energy) to accommodate growing economic activity in the manufacturing and services sectors. Moreover, local governments could redouble their efforts to deal with the pollution problems that plague many Indian cities. Government interventions in urban centres could also include improvements in delivering education, housing and other urban services. Local governments could also create social networks and support systems to help migrants integrate into local labour markets.

Information could be disseminated to transition-zone residents on how they could improve their livelihoods, health and education by migrating to more urbanized areas. Such information could highlight research showing the positive effects of rural–urban migration on income (e.g. Young, 2013). Information dissemination campaigns could also be used to discourage potential migrants from entering the transition zone. In-migration cannot be stopped by government edict, but it can be discouraged by providing information to potential migrants using a mass education programme regarding the dangers associated with life in the transition zone. Potential in-migrants could be reached by disseminating information in locations known (from the 2011 household survey) to be common departure points for in-migrants.

Social scientists have not yet developed complete explanations for why people migrate from rural to urban areas (Bryan, Chowdhury, & Mobarak, 2014). Thus, implementation of elements of the voluntary out-migration programme we propose should be treated as small-scale experiments designed such that data can be gathered to assess effectiveness. Results can be used to modify subsequent efforts to influence out-migration decisions.

In the long term, residents in the transition zone will leave, because of the challenges associated with living there, and this may offer opportunities to augment the mangrove forest. An expanded forest will provide enhanced opportunities for conserving biodiversity and serve as a bio-shield to protect Kolkata and other inland cities from the high winds associated with cyclonic storms over the Bay of Bengal.

The proposed out-migration approach would require a relatively long time frame, during which those living in the transition zone would face increasing risks. Thus, during this interval, there is a need to reduce potential threats by undertaking embankment realignment and upgrading, and mangrove bio-shield restoration, as well as strengthening early-warning systems, emergency preparedness and cyclone shelter investments to more effectively address the treatment of disasters when they do occur. These measures are complementary with initiatives that encourage out-migration, and would be phased out as the population in the most exposed areas of the transition zone decreases.

Conserving biodiversity

One element of the proposed approach to conserving biodiversity concerns improved coordination of forest management procedures between India and Bangladesh. Although the study concerned the Indian portion of the Sundarbans, the study team recommended that the Sundarbans ecosystem be managed by having both Bangladesh and India create an integrated policy for conservation and development, or at least adopt a harmonized policy that could be implemented by each country individually. Steps towards bilateral cooperation on biodiversity conservation were advanced in 2011 when the governments of India and Bangladesh entered into cooperative agreements that facilitate information sharing and increased coordination in the context of managing the Sundarbans forest (Ortolano et al., 2016).

Biodiversity conservation could also be furthered by the previously mentioned activities involving mangrove restoration in new land areas created by realigning existing embankments to prevent embankment failure due to tidal creek erosion. Other measures aimed at conserving biodiversity involve generating revenues to benefit local residents, giving them a stake in forest conservation (Biller & Sanchez-Triana, 2013). One approach to revenue generation involves payment for ecosystem services (Pattanayak, Wunder, & Ferraro, 2010). Services provided could include the previously mentioned use of the forest as a bio-shield to protect Kolkata and other inland areas from damaging winds. Also, an innovative revenue-generation approach that can be further explored involves creating property rights on non-timber forest products, such as honey, fruit and other extracted products and non-extractive use functions (e.g. tiger viewing). Moreover, by preserving the mangrove it will be possible to access revenue streams that have been (and will be) created as part of carbon financing programmes intended to manage greenhouse-gas emissions. New revenues generated in the aforementioned ways could provide benefits for residents, such as programmes to provide transition-zone residents the training needed to access jobs in the stable zone and outside the Sundarbans.

Another biodiversity-conservation measure involves establishing privately funded hatcheries for use in shrimp aquaculture and improving shrimp seed collection methods while hatcheries are being developed. Aquaculture in the Sundarbans could be made more sustainable if state-of-the-art hatcheries were promoted and if the Food and Agricultural Organization's Code of Conduct for Responsible Fisheries were adopted.

Institutions

Government programmes for the region's development are driven largely by agencies in the government of West Bengal. The Department of Sundarban Affairs and the Forest Department have the broadest administrative influence over the Sundarbans. When the Department of Sundarban Affairs was created in 1994, it subsumed the SDB, which had been created to address the socio-economic challenges of the region. A report by World Wildlife Fund clarifies the political challenges faced in planning for the region:

> The original mandate of SDB was to coordinate activities of the various sectoral departments operating in the ecoregion. However, the SDB has since morphed into a parallel implementation agency with a range of divisions mirroring the departmental mandates of different state departments. (Danda, Sriskanthan, Ghosh, Bandyopadhyay, & Hazra, 2011, p. 22)

The number of poorly adapted initiatives that have been used in the Sundarbans over the past decades makes it evident that there are no effective incentives for government agencies to cooperate or coordinate their programmes in the region. Numerous examples exist. For instance, aquaculture development has contributed to tidal creek erosion that undermines embankment stability. Another example is the complete ban on mangrove cutting, which reduces the incentives for communities to manage them well and has in fact turned the pursuit of some potentially sustainable uses of the forest into a crime.

Although the Department of Sundarban Affairs is intended to coordinate all efforts in the region, tackling the high-priority issues could be improved by enhancing the cooperation and coordination between selected pairs of responsible departments of the government of West Bengal. For example, cooperation between the Forest Department and the Irrigation and Waterways Department is essential given the current and future need to encourage mangrove growth along areas that become available for restoration as a result of embankment realignment. There is a need to ensure that areas created by retreat of embankments are immediately planted with appropriate mangrove species and maintained. This requires the Irrigation and Waterways Department to (a) plan and implement the embankment realignment activities sequentially, and engage in mangrove planting at the time embankments are retreated; and (b) hand over jurisdiction of land vacated as embankments are realigned to the Forest Department without any opportunity for other land uses to occupy that land. The Department of Panchayat and Rural Development should also be a part of an enhanced coordination effort, given its capacity to improve community efforts in social forestry.

Another example involves coordination between the Irrigation and Waterways Department and the Department of Fisheries and Aquaculture. Much of the current embankment erosion in the eastern delta is related to the location of aquaculture ponds and the timing of pond refilling activities. The Department of Fisheries and Aquaculture should promote aquaculture practices that do not negatively impact embankment investments by the Irrigation and Waterways Department. Enhanced cooperation and communication between these two departments could reduce the undesirable effects associated with the location and operation of aquaculture activities. For example, both departments should take steps to prohibit pond operators from operating their sluices at the same time to let water in to flush the ponds on the high spring tides, because this causes a major discharge.

The study team presented its recommended strategy to officials in the governments of India and West Bengal, and the recommendations are feeding into meetings and consultations aimed at arriving at a consensus among stakeholders regarding next steps in managing the Sundarbans. This consensus-building process continues, and thus complete government policies for dealing with the water-related issues described herein are still being formulated.

Definitive policy making is going on in a number of domains, such as promoting transboundary management of protected areas, strengthening the mandate of the SDB, and augmenting the budgetary resources for conservation and improvement of community livelihood options. In addition, many of the study team's recommended actions are already being pursued. As an example, the Irrigation and Waterways Department's programme of embankment reconstruction has been modified in ways consistent with the study team's geomorphological modelling results. The department's USD 1 billion of funding from the government of India is being used to realign damaged embankments in the transition zone. Other examples include: construction of 200 cyclone shelters under World Bank–financed projects; construction of additional cyclone shelters using financing from NGOs; and mangrove planting by private-sector actors to earn carbon credits under the Clean Development Mechanism of the UN Framework Convention on Climate Change. These activities rely directly or indirectly on local capacity, which has been informed or strengthened by the study team's efforts.

Conclusions

Persistent pressures from water-related threats – SLR, soil and water salinization, and flooding due to embankment overtopping and failure – have made the Indian Sundarbans a challenging place to live, and effects of global climate change will only worsen conditions. Attempts at poverty alleviation in high-risk areas will involve an endless and unavailing struggle, even in the absence of future climate change impacts.

The study team's recommended approach involves striking an effective balance between (a) long-term risk avoidance via voluntary out-migration and (b) risk-reduction measures for those who remain in the high-risk transition zone. As mentioned, this combination of approaches is not internally contradictory once the time scales involved are considered. Time will be required to equip those who choose to out-migrate with the economic, human and social capital needed to migrate successfully to safer areas with greater opportunities. Measures to assist with the out-migration include: improve interagency collaboration; invest in key areas, such as the creation and operation of job training centres for migrants; augment services in nearby urbanized areas in the stable zone and beyond so that those areas are better able to receive migrants; and implement incentives to encourage out-migration (e.g. information dissemination campaigns to alert residents regarding the increasing risks of remaining and the improved options elsewhere) and to discourage in-migrants.

During the time needed to prepare for successful voluntary out-migration, measures to cut risks (e.g. embankment re-alignment and enhanced disaster-management systems) will be needed, because the population will continue to be exposed to considerable risk.

A number of additional recommendations concern biodiversity conservation (e.g. creation of revenue streams to give residents living near the SRF incentives to conserve the forest).

Acknowledgements

The findings, interpretations, and conclusions expressed in this article do not necessarily reflect the views of the executive directors of the World Bank or the governments they represent. The World Bank does not guarantee the accuracy of the data included in this work. The boundaries, colours, denominations and other information shown on any map in this work do not imply any judgment on the part of the World Bank concerning the legal status of any territory or the endorsement or acceptance of such boundaries.

Many thanks are due to members of the study team. In addition to the authors, the study team included Anupam Joshi, Cecilia Belita, Charles Joseph Cormier, Dan Biller, Ghazal Dezfuli, Jack Ruitenbeek, John Pethick, Jean-Noel Guillossou, Nalin Jena, Priti Kumar, Rahul Kanakia, Ana Luisa Gomes, Ramachandran R. Mohan, Roshni Sarah John, Ruma Tavorath, Sanjay Gupta, Santiago Enriquez, Shakil Ahmed Ferdausi, Snehashish Raichowdhury, Soumya Kapoor and Surbhi Goyal. Additional background papers and analyses were produced by Anurag Danda, Anna C. O'Donnell, Bjorn Larsen, Chandreyee Das, Christine Heyniger, Damanjit Singh Minhas, D. M. Mohan, Elena Strukova Golub, Gautam Gupta, Ishtiaque Ahmed, Juan Luna-Kelser, Kakoli Sen Sarma, Lidvard Gronnevet, Nirmala Banerjee, Peter Webster, Pranabes Sanyal, Priyanka Roy, Ratul Saha, Samir Mehta, Samuel Taffesse, S. C. Rajshekar, Somenath Battacharya, Subimal Roy, the Institute of Environmental Studies and Wetland Management, Economic Information Technology, IMRB Consultants, Jadavpur University, Sundarban Rural Development and Training Centre, West Bengal Green Energy Department Corporation and the Worldwide Fund for Nature–India. We are also grateful to Dinesh Kumar, Cecilia Tortajada, John Pethick and two anonymous reviewers for helpful comments on an earlier draft.

Disclosure statement

No potential conflict of interest was reported by the authors.

Funding

This work is funded by the Australian Agency for International Development [SAWI Trust Fund], the Department for International Development [SAWI Trust Fund], and the World Bank [SAWI Trust Fund].

References

Allison, M. A. (1998). Historical changes in the Ganges-Brahmaputra delta front. *Journal of Coastal Research, 14*, 1269–1275.

Antonov, J. I., Levitus, S., & Boyer, T. P. (2002). Steric sea level variations during 1957–1994: Importance of salinity. *Journal of Geophysical Research: Oceans (1978–2012), 107*(C12), SRF-14.

Auerbach, L. W., Goodbred, S. L., Jr, Mondal, D. R., Wilson, C. A., Ahmed, K. R., Roy, K., & Ackerly, B. A. (2015). Flood risk of natural and embanked landscapes on the Ganges-Brahmaputra tidal delta plain. *Nature Climate Change, 5*, 153–157.

Banerjee, K. (2013). Decadal change in the surface water salinity profile of Indian Sundarbans: A potential indicator of climate change. *J Marine Sci Res Development, S11*, 002.

Banerjee, K., Roy Chowdhury, M., Sengupta, K., Sett, S., & Mitra, A. (2012a). Influence of anthropogenic and natural factors on the mangrove soil of Indian Sundarbans wetland. *Arch Environ Sci, 6*, 80–91.

Banerjee, K., Senthilkumar, B., Purvaja, R., & Ramesh, R. (2012b). Sedimentation and trace metal distribution in selected locations of Sundarbans mangroves and Hooghly estuary, Northeast coast of India. *Environmental Geochemistry and Health, 34*, 27–42.

Basu, S. (2010). A shelter too far for cyclone & flood relief, The Statesman. Retrieved January 10, 2012, from http://www.thestatesman.net/index.php?option=com_content&view=article&id=328601&catid=42

Bhattacharyya, S., Pethick, J., & Sarma, K. S. (2013). Managerial response to sea level rise in the tidal estuaries of the Indian Sundarbans: A geomorphological approach. *Water Policy, 15*, 51–74.

Biller, D., & Sanchez-Triana, E. (2013). Enlisting markets in the conservation and sustainable use of biodiversity in South Asia's Sundarbans. *International Journal of Social Ecology and Sustainable Development, 4*, 71–86.

Bryan, G., Chowdhury, S., & Mobarak, M. (2014). Underinvestment in a profitable technology: The case of seasonal migration in Bangladesh. *Econometrica, 82*, 1671–1748.

Chakraborty, S. (2015). Investigating the impact of severe cyclone Aila and the role of disaster management department - A study of Kultali block of Sundarban. *American Journal of Theoretical and Applied Business, 1*, 6–13.

Chatterjee, M., Shankar, D., Sen, G. K., Sanyal, P., Sundar, D., Michael, G. S., ... Sarkar, K. (2013). Tidal variations in the Sundarbans Estuarine system. *Journal of Earth System Science, 122*, 899–933.

Danda, A. A. (2007). *Surviving in the Sundarbans: Threats and responses* (Unpublished PhD). University of Twente, The Netherlands.

Danda, A. A. (Ed.). (2010). *Sundarbans: Future imperfect climate adaptation report*. Kolkata: WWF-India.

Danda, A. A., Sriskanthan, G., Ghosh, A., Bandyopadhyay, J., & Hazra, S. (2011). *Indian Sundarbans delta: A vision*. New Delhi: World Wildlife Fund for Nature-India.

Das, C. S. (2014). Causes, consequences and cost-benefit analysis of the conflicts caused by tiger straying incidents in Sundarban, India. In *Proceedings of the Zoological Society* (pp. 1–11). New Delhi: Springer India.

Eaton, S. (2015). Climate change-linked 'everyday disasters' are displacing the world's poorest people. *PRI's The World*, December 03, 2015. Retrieved May 4, 2016, from http://www.pri.org/stories/2015-12-03/climate-change-linked-everyday-disasters-are-displacing-world-s-poorest-people

Ericson, J. P., Vörösmarty, C. J., Dingman, S. L., Ward, L. G., & Meybeck, M. (2006). Effective sea-level rise and deltas: Causes of change and human dimension implications. *Global and Planetary Change, 50*, 63–82.

Ghosh, A. (2015). Everyday disasters' driving flight from Sundarbans. *Rueuters*, Apr 7. Retrieved April 20, 2015, from http://uk.reuters.com/article/2015/04/07/uk-india-sundarbans-migration-idUKKBN0MY0B220150407

Ghosh, A., Schmidt, S., Fickert, T., & Nüsser, M. (2015). The Indian Sundarban mangrove forests: History, utilization, conservation strategies and local perception. *Diversity, 7*, 149–169.

Giri, C., Pengra, B., Zhu, Z., Singh, A., & Tieszen, L. L. (2007). Monitoring mangrove forest dynamics of the Sundarbans in Bangladesh and India using multi-temporal satellite data from 1973 to 2000. *Estuarine, Coastal and Shelf Science, 73*, 91–100.

Glaeser, E. A. (2011). *Triumph of the city*. New York: Penguin Press.

Guzzella, L., Roscioli, C., Viganò, L., Saha, M., Sarkar, S. K., & Bhattacharya, A. (2005). Evaluation of the concentration of HCH, DDT, HCB, PCB and PAH in the sediments along the lower stretch of Hugli estuary, West Bengal, Northeast India. *Environment International, 31*, 523–534.

Hazra, S., Ghosh, T., DasGupta, R., & Sen, G. (2002). Sea level and associated changes in the Sundarbans. *Science and Culture, 68*, 309–321.

Hazra, S., Samanta, K. Mukhopadhyay, A., & Akhand, A. (2010). *Temporal change detection (2001-2008) study of Sundarban*. Jadavpur, Kolkata, India: School of Oceanographic Studies, Jadavpur University. Retrieved December 7, 2015, from http://www.iczmpwb.org/main/pdf/ebooks/WWF_FinalReportPDF.pdf

Kay, S., Caesar, J., Wolf, J., Bricheno, L., Nicholls, R. J., Islam, A. S., & Lowe, J. A. (2015). Modelling the increased frequency of extreme sea levels in the Ganges–Brahmaputra–Meghna delta due to sea level rise and other effects of climate change. *Environmental Science: Processes & Impacts, 17*, 1311–1322.

Knowler, D., Philcox, N., Nathan, S., Delamare, W., Haider, W., & Gupta, K. (2009). Assessing prospects for shrimp culture in the Indian Sundarbans: A combined simulation modelling and choice experiment approach. *Marine Policy, 33*, 613–623.

McKinsey Global Institute. (2010). India's urban awakening: Building inclusive cities, sustaining economic growth. Retrieved April 22, 2015, from http://www.mckinsey.com/insights/urbanization/urban_awakening_in_india

Narayan, D. Pritchett, L., & Kapoor, S. (2009). *Moving out of poverty, volume 2, success from the bottom up*. Washington, DC: The World Bank.

National Disaster Management Authority. (2015). Environment and social management framework, National Cyclone Risk Mitigation Project II, National Disaster Management Authority, Government of India, Kolkata. Retrieved December 13, 2015, from http://www-wds.worldbank.org/external/default/WDSContentServer/WDSP/IB/2015/03/19/000442464_20150319090337/Rendered/PDF/E45560V10REVIS00Box385436B00PUBLIC0.pdf

Ortolano, L., Sánchez-Triana, E., Paul, T., & Ferdausi, S. A. (2016). Managing the Sundarbans region: Opportunities for mutual gain by India and Bangladesh. *International Journal of Environment and Sustainable Development, 15*, 16–31.

Pattanayak, S. K., Wunder, S., & Ferraro, P. J. (2010). Show me the money: Do payments supply environmental services in developing countries? *Review of Environmental Economics and Policy, 4*, 254–274.

Penning-Rowsell, E. C., Sultana, P., & Thompson, P. M. (2013). The 'Last resort'? Population movement in response to climate-related hazards in Bangladesh. *Environmental Science & Policy, 27*, S44–S59.

Pethick, J., & Orford, J. D. (2013). Rapid rise in effective sea-level in southwest Bangladesh: Its causes and contemporary rates. *Global and Planetary Change, 111*, 237–245.

Raha, A. K., Zaman, S., Sengupta, K., Bhattacharyya, S. B., Raha, S., Banerjee, K., & Mitra, A. (2013). Climate change and sustainable livelihood programme: A case study from Indian Sundarbans. *J Ecol, 107*, 335–348.

Ravallion, M. (2007). Urban poverty. *Finance and Development, 44*, 15–17.

Samanta, S., Mitra, K., Chandra, K., Saha, K., Bandopadhyay, S., & Ghosh, A. (2005). Heavy metals in water of the rivers Hooghly and Haldi at Haldia and their impact on fish. *Journal of environmental biology/Academy of Environmental Biology, India, 26*, 517–523.

Sánchez-Triana, E., Paul, T., Ortolano, L., & Ruitenbeek, J. (2014). *Building resilience for sustainable development for the West Bengal Sundarbans – Strategy report* (Report No. 88061-IN). Washington, DC: World Bank. Retrieved April 6, 2014, from http://documents.worldbank.org/curated/en/2014/01/20162806/building-resilience-sustainable-development-sundarbans-strategy-report

Sinha, M., Mukhopadhyay, M. K., & Mitra, P. M. (1997). Impact of Farakka Barrage on the hydrology and fishery of Hooghly Estuary. *Oceanographic Literature Review, 4*, 397.

Wallingford, H. R., ABPmer, & Pethick, J. (2006). *Review and formalisation of geomorphological concepts and approaches for estuaries, R&D Technical Report FD2116/TR2*. London: Defra.

Woodroffe, C. D., Nicholls, R. J., Saito, Y., Chen, Z., & Goodbred, S. L. (2006). Landscape variability and the response of Asian megadeltas to environmental change. In N. Harvey (Ed.), *Global change and integrated coastal management* (Vol. 10, pp. 277–314). Dordrecht, The Netherlands: Springer.

Young, A. (2013). Inequality, the urban-rural gap and migration. *The Quarterly Journal of Economics, 128*, 1727–1785.

Techno-institutional models for managing water quality in rural areas: case studies from Andhra Pradesh, India

V. Ratna Reddy

ABSTRACT
This article examines the rationale, technologies, economics and institutional modalities in water quality management operations to draw lessons for designing policies for sustainable service delivery at scale. While the rationale for providing potable drinking water at affordable prices is clear, their economic viability is weak given their present scale of operations. There is a need for institutional safeguards for selection of deserving villages and water quality monitoring. It is argued that public–private–community partnerships are economically viable and sustainable. Adopting appropriate technologies could help with addressing the water quality issues in a more comprehensive manner.

Introduction

The quality of drinking water in rural areas is a key policy concern for the government of India. It is estimated that access to drinking water in rural areas is as high as 90% at the household level, but a significant proportion of these households still have to manage with poor-quality water. Bacteriological contamination due to poor sanitary conditions remains the major cause of concern. Other widely reported sources of contamination include: fluoride, which is prevalent in 18 states, affecting 25 million people; arsenic, which affects 62 million people in 16 states; and intensive use of fertilizers and pesticides in agriculture, which also results in contamination of surface water bodies (through runoff) and groundwater (through seepage) (Government of India [GoI], 2006; Wate, 2012). The incidence and intensity of drinking water contamination and pollution are higher among poor households. Provision of safe drinking water in a sustainable manner, therefore, is crucial for improved quality of life in rural areas in general and poor households in particular.

In the united Andhra Pradesh, access to potable water in rural households is marginally better (73%) than the national average of 69% for rural areas. Andhra Pradesh is among the worst fluoride-affected states in the country, with an estimated 1.88 million habitations (cluster of dwellings within a village where people live, similar to a hamlet) reporting fluoride incidence. Although there has been a considerable decline in the number of affected habitations during the last decade, the magnitude of the problem remains high. Andhra Pradesh also has a high incidence of waterborne diseases, especially in the coastal regions. Jaundice morbidity is on the higher side, at 16,110 per million persons, against 14,100 per million at the all-India level, while diarrhoea-related morbidity in children younger than 3 is on the lower side at 15%, against 19% at the all-India level (Pushpangadan, 2006). In the coastal areas of Andhra Pradesh, 500,000 people live in pockets where severe faecal contamination of drinking water sources is above 2000 most probable number/ml. People also suffer from water-washed diseases, i.e. diseases caused by physical contact with contaminated water, such as scabies and trachoma (Reddy & Kurian, 2010).

This is the case despite a number of government initiatives towards providing safe drinking water in rural areas. The state government–initiated community-based programmes face challenges due to lack of local participation in planning, implementation and management, and the lack of clarity on the role of the local government (the *panchayti raj*, the smallest institution of local self-government in India, administers a few villages under it) in supporting local initiatives. On the other hand, there is growing evidence that some of the public initiatives involving private agencies are effective in addressing drinking water quality issues in the rural areas. Similarly, there are non-governmental organizations working to address this issue effectively. There are also individual initiatives. The evidence on the effectiveness of public–private partnership initiatives is rather mixed, with both positive and negative outcomes reported (Athena Infonomics, 2012; Gopakumar, 2010; Naiga, Penker, & Hogl, 2015; Lloyd Owen, 2016; van Wijk-Sijbesma & van Dijk, 2006). The operational modalities are, however, not uniform across agencies and partners. There is a need to assess the approaches and techno-institutional models adopted by various agencies in order to identify the ones which have the potential for scaling up at the national level.

Objectives and scope

This article aims to identify the features of sustainable models and systems that are capable of addressing drinking water quality problems in rural areas and that could be replicated at scale through systematic investigation. Learning from such an investigation can be used as a framework for identifying economically viable, environmentally friendly and socially acceptable (providing inclusiveness and equity) technical solutions. This article attempts to find out the most important characteristics of techno-institutional models that sustain water supply programmes in the long run. The learning based on the case studies is expected to facilitate better policy design across the country. Specific objectives include: scoping the various initiatives; understanding what has worked and what has not; and identifying the critical inputs required for scaling up and scaling out feasible options.

Several agencies are involved in providing clean water for drinking purposes in rural areas, wherein private–public–community partnerships are forged. In the present study four different technologies appropriate to the study regions are chosen. These technologies were adopted depending on the type of water contamination and the costs involved. Different institutional arrangements were adopted by the implementing agencies, though community involvement and participation were common to all the models. The study takes a comprehensive view and covers economic, social and institutional aspects of the initiatives.

The study also covers different agro-climatic regions and resource endowments, which include low-rainfall regions that depend entirely on groundwater and medium-rainfall regions endowed with surface water supplies, especially canal water supply. The nature of contamination and the technology adopted for water purification vary between the regions.

Sample villages are equally distributed between the two sources of drinking water: surface and groundwater. The sample sites are representative of arid and semi-arid regions, including rainfed areas and canal commands, which account for 50% of the rural population. Thus, the findings of the study will have wide policy relevance.

Approach and methodology

The study was carried out using qualitative and quantitative data at three levels: agency, village / treatment plant and household. As the first step, the agencies involved in the supply of treated water were identified. As the second step, based on the information provided by various agencies, the major players in the supply of treated water using different technologies were identified. In addition to the number of villages covered over the years, the agencies were asked to list the best five villages where the initiatives are running successfully and had been handed over to the village community or to the village panchayat.

As the third step, 8 villages from a list of 20 were selected. Two criteria were used to select the sample villages: villages where water purification plants are managed by the local community and village panchayat; and old and well-performing initiatives. Based on our discussions with the officials in the Rural Water Supply and Sanitation Department (RWSSD) and interactions with various agencies, four agencies (Byrraju Foundation, Naandi Foundation, Sai Oral Health Foundation and Rural Economic and Educational Development Society [REEDS]) were identified, in consultation with RWSSD, as the potential agencies for the study. The rationale for selecting these agencies is that they are operating in number of villages and are effective in addressing the water quality issues in different regions. They also adopt different economic and institutional modalities, providing wide-ranging experience. The Byrraju Foundation and the Naandi Foundation are supported by the corporate sector as part of their corporate social responsibility work; the Sai Oral Health Foundation is a non-governmental organization; and REEDS operates in collaboration with RWSSD.

For detailed analysis, two villages from the Byrraju Foundation and three villages from the Naandi Foundation were selected. All five sample villages are using ultraviolet (UV) and reverse osmosis (RO) technologies. UV technology is adopted in the villages where faecal contamination is prevalent, and RO is used to purify water with high fluoride concentration. One village from the Sai Oral Health Foundation, which is involved in defluoridation at the community level, was selected. The Sai Oral Health Foundation adopted a low-cost bone char technology to purify fluoride-contaminated water. Finally, two sample villages, one covered by RWSSD alone and another in collaboration with REEDS, were selected. Microfilter technology was adopted in one of the villages, and RO technology is used in the other. Care was taken to cover all the technologies in use in the region. The eight sample villages are spread over eight *mandal*s (sub-district administrative regions) in four districts across the Coastal and Telangana regions of united Andhra Pradesh (Table 1). Four different technologies are covered, though UV and RO are the most widely used.

Both qualitative and quantitative information was elicited for this purpose. Focus group discussions with stakeholders and village transect walks were conducted to elicit qualitative information. Based on the qualitative information, detailed questionnaires were prepared to obtain information at the treatment plant and household levels. All the households in the village were grouped by land ownership, which is a proxy for economic status in rural areas. A sample of 30 households from each village was selected using representative and

Table 1. Sample Villages by Agency and Technology (Andhra Pradesh, India).

Sample village	Mandal	District	No. of households	Agency	Technology
Lachammaguda	Narayanpur	Nalgonda	600	SOHF	Bone char
Dandu Malkapur	Choutappal	Nalgonda	1,032	RWSSD	RO
Kasaram	Yadagirigutta	Nalgonda	261	Naandi	RO
Kandlakoya	Medak	Rangareddy	507	Byarraju	RO
Bomminampadu	Mudinepally	Krishna	500	Naandi	UV
Bhairavapatnam	Mandavalli	Krishna	303	REEDS/ RWSSD	Microfilter
Telaprolu	Unguturu	Krishna	3,000	Naandi	UV
Puduru	Puduru	West Godavari	900	Byarraju	UV

Note: SOHF= Sai Oral Health Foundation; RO = reverse osmosis; RWSSD = Rural Water Supply and Sanitation Department; UV = ultraviolet. Source: Field survey, calculated by the author.

proportionate sampling covering all the important socio-economic groups, including households that do not buy water from the water purification plants. On the whole, detailed information was collected from 240 households in eight villages.

The broad profile and characteristics of the sample households reflect heterogeneity within the village and homogeneity of the communities across the sample sites. Within each sample village, a representative sample of different socio-economic groups and also the buyers and non-buyers of water was drawn. Similar sample profiles were adopted across the sites, though the socio-economic profile of the sample households may not be the same across the sample villages, due to the inter-village variations in socio-economic structure. But the relative standing of each socio-economic group within a village is expected to be same across the villages, i.e., a poor scheduled caste household (better-off large farming household) is expected to face similar disadvantages (advantages) across the villages.

Rationale for provision of treated water in the sample villages

Water contamination was widespread, irrespective of the source of water. In all the sample villages, raw water samples from wells contained fluoride in excess of permissible limits (Table 2). In the case of Bomminampadu Village, which also depends on groundwater, the problem was of water hardness due to the intrusion of seawater. Surface water resources were contaminated bacteriologically and chemically, causing waterborne as well as water-washed diseases.

The health impacts of fluoride contamination are long-term in nature and cause physical disabilities. In the case of contamination of water with faecal matter, the impacts range from short- to medium-term, in the form of diseases such as diarrhoea, cholera, joint pain, etc. A related study in the study region found that the extent of health impacts ranges between 21% and 57% of the sample households due to faecal contamination (Reddy, Kullappa, & Rao, 2008). Health impacts are often measured in terms of number of persons and number of days of sickness. More importantly, the incidence of diseases is higher among poor households (below the poverty line) than among non-poor households (above the poverty line). Women are more affected than men by exposure to contaminated water because women are the main fetchers and users of water. Health impacts result in economic impacts, as households incur expenses for treatment (medicine and doctor visits) and also forgo wage earnings, when sick persons cannot work. The economic losses due to health impacts are USD 8.5–25 per household (2008). These costs are much higher in the case of fluoride contamination, due to the long-term nature of the impacts, which include permanent

Table 2. Quality of Raw Water in the Sample Villages in Andhra Pradesh, India.

Parameter (permissible)	Lachammaguda	Dandu Malkapur	Kasaram	Kandlakoya	Puduru	Bomminampadu	Telaprolu	Bhairavapatnam
pH (6.5–8.5)	---	---	7.49	7.21	7.73	6.75	7.97	8.68
Electrical conductivity (750 mho)	---	---	620	1,715	---	409	580	818
Dissolved solids (ppm<500)	---	---	1,000	1,128	160	254	370	506
Total hardness as $CaCO_3$ (ppm<300)	---	---	664	824	108	102	148.32	204
Alkalinity to methyl orange (ppm<200)	---	---	740	260	113	33.2	180.6	224
Chlorides (ppm<352)	---	---	553.8	320	46.99	54.6	77.76	136
Sulphate (ppm<208)	---	---	273.6	255	---	5.1	---	14
Nitrate (ppm<38)	---	---	47	24	50.10	0.05	---	None
Fluoride (ppm<1.00)	4.2	1 to 2	1.24	1.56	---	0.68	---	None
Iron (ppm<0.3)	---	---	None	0.02	---	0.03	---	None
Turbidity (ppm<5.0)	---	---	---	2.6	10	0.49	2	11.0
E. coli (<50/100 ml) / coliform (<10)	n/a	n/a	n/a	n/a	30,000	9/129	None	1,100

Notes: Figures in brackets indicate permissible limits; n/a = not applicable; --- = data not available; None = Not present.
Source: Respective water treatment plants.

physical disability. This emphasizes the necessity of providing potable drinking water in the region.

There are a number of water treatment technologies in operation in the villages. They range from simple and 'easy to manage' to very sophisticated. While RWSSD provided slow sand filters, sophisticated technologies such as RO and UV are being promoted by corporate and private agencies. RO and UV are compact and effective. These technologies are established and managed by these agencies on commercial basis. Their packaging, marketing and promotion seem to have caught the imagination of communities in some villages, where buying water from the treatment plant has become a status symbol among middle- and upper-middle-class households.

The detailed survey of the sample villages reveals that communities face problems associated with poor quality of drinking water, though the intensity and extent of the problems vary between surface and groundwater sources. Health problems were more serious in the case of groundwater-based sources affected by fluoride. Hence, the demand for treated water was higher in the groundwater-dependent areas than in the surface water–dependent areas. Of the sample villages, Lachammaguda was the worst-affected in terms of poor access to potable water. On the other hand, one of the sample villages, where water quality is permissible, was fully covered by the Naandi Foundation. This indicates that there is commercial interest in setting up water purification plants. In the villages which were dependent on contaminated water from surface sources, the demand for treated water was limited, as people do not seem to be convinced of the benefits of clean water. But in Bhiravapatnam, with its functioning water purification plant (using slow sand filters) and distribution system, the demand for treated (microfiltered) water was reasonably high (66%), due to high awareness coupled with free supply. Bomminampadu has also high demand (64%) for treated water due to the hardness of the raw water.

It may be noted that the UV and RO plants have a scale limitation: the demand for water is restricted to drinking and cooking needs (about 10% of the total domestic water requirement), owing to lack of affordability. As a result, these models are effective only in controlling waterborne diseases, while the problems associated with water-washed diseases continue. Communities are exposed to water-washed diseases when they use contaminated water for bathing, cleaning and washing clothes. On the other hand, microfilters can meet the entire household water demand at low cost and can replace slow sand filters in rural areas that are not affected by fluoride contamination.

Overview of the initiatives

The main features of the models followed by the selected agencies in terms of the technologies, operational modalities, coverage and economics are discussed in the subsequent paragraphs. They are also summarized in Table 3.

Byrraju Foundation

This non-profit organization, set up in July 2001, aims to improve quality of life in the villages. The foundation lays emphasis on providing sustainable safe drinking water to villagers through a low-cost community-driven model, integrating technology and community participation. To address this issue, the foundation has taken up the project Safe Water for

Table 3. Overview of Important Agencies Involved in Water Purification in Rural Andhra Pradesh, India.

	Agency			
Details	BF	NF	SOHF	REEDS/RWSSD
Scale of operations	1 state, 5 districts, 56 plants (about 188 villages; 950,000 population	2 states, 13 districts, 123 plants (villages), 1,450,000 population	1 state, 1 district, 22 plants, 5,000 population	140 gram panchayats, 100,000 population
Technology	RO in groundwater regions (fluoride and mineral contamination); UV in surface water regions (bacterial contamination)	RO in Groundwater regions (fluoride and mineral contamination); UV in surface water regions (bacterial contamination)	Bone char; deflouridation in fluoride-affected regions	Microfilter
Scalability	Low	Low	Low	High
Break-even size	3–4 months at 30% capacity utilization (1,000–2,000 lph capacity)	--	Very low cost; 3–4 paise per litre	Very low cost; 0.02 paise per litre
Finance	Panchayat/community and BF	Gov. of India / state gov. / philanthropists / institutional finance (CAPEX)	Foundation/ community	RWSSD/community
User fee	12.5 paise per litre	12.5 paise per litre	Free	RWSSD / PRI norms / free
Institutional arrangements	Public–panchayat–private (NGO) partnership	Panchayat–private (NGO) partnership	Community	Panchayat/ community

Notes: BF = Byrraju Foundation; NF = Naandi Foundation; SOHF = Sai Oral Health Foundation; REEDS = Rural Economic and Educational Development Society; RWSSD = Rural Water Supply and Sanitation Department; CAPEX = Capital Expenditure; PRI= Panchayati Raj Institution; gram panchayat = village council; paise = 1/100 of rupee. Source: Gathered by the author from the respective agencies.

Everyone Using Effective Technology, conforming to World Health Organization guidelines and the norms of the Bureau of Indian Standards.

The foundation's vision is that all processes are owned, managed and led by the community, and facilitated by the village development council (*gram vikasa samithi*), a voluntary body with nine members, each one looking after a particular initiative. The foundation has established 56 drinking water purification plants in five districts of Andhra Pradesh. The technology is chosen based on the source of water and the nature of contamination. In the case of bacteriological contamination of water, UV is used, and in the case of chemical contamination (such as high fluoride or salinity), RO is used. Treated water is supplied in food-grade cans at the plant at USD 0.022 for a 12-litre can. It is delivered to the house for an additional charge (USD 0.0074–0.029 per can, depending on the distance).

The cost of the purification plants ranges from INR 0.75 million for UV to USD 12,700 for RO plants. Of this, 80% of the cost is borne by the community and the rest by the foundation (for land, USD 750–1500 is allotted free by the panchayat; for the building, 500 sq. feet, at USD 5970, is borne by the village community; for equipment, USD 3730 for UV, or USD 5220 for RO, is shared equally by the community and the foundation; for 1000 water cans, USD 1090 is 50% borne by consumers). In the case of the Byrraju Foundation, the plant does not get power at a concessional rate.

Naandi Foundation

The Naandi Foundation was established in 1998 as an autonomous, non-profit trust dedicated to changing the lives of the marginalized groups in India through efficient public–private partnerships. Naandi is working with about two million people, in some of the most challenging terrains in eight states and one union territory. Naandi has partnered with Water Health India and TATA Projects Limited to address the problem of drinking water contamination in rural areas. Water Health India employs a patented non-submersible UV-based water treatment process; TATA provides RO technology to treat chemically contaminated water.

The capacity of the UV plants is 22,000–66,000 litres per day, while that of the RO plants is 11,000–30,000 litres per day. The initial capital for setting up the community-based safe drinking water system is provided by a variety of stakeholders, including governments, corporations, private foundations, philanthropists, non-resident Indians, national and international institutional donors and the communities which represent the potential end users. The cost ranges from USD 10,000 (for 12,000 litres per day) to USD 13,600 (for 20,000 litres per day) for RO technology; USD 21,640 in the case of UV technology. End users pay USD 0.016 for 12 litres and USD 0.20 for 20 litres of safe drinking water from UV plants and USD 0.20 for 20 litres from RO plants, to ensure that the majority of village households can afford it.

Sai Oral Health Foundation

This organization was established in 1987 as a service-oriented, non-profit, voluntary organization dedicated to ensuring health for all poor and needy people. The foundation has adopted innovative approaches to reduce fluoride levels in drinking water.

A bone char–based fluoride filtration system is being used at the community level. Processing capacity varies depending on the concentration of fluoride, from 100,000 litres per day when the fluoride concentration is below 2 mg/litre to 50,000–35,000 litres when fluoride concentration is 4–5 mg/litre. On average, defluoridation of 1 m^3 of water with a fluoride concentration of 3–4 mg/litre costs USD 0.42–0.59. A community-level defluoridation filter (bone char technology) costs about USD 224, with a running cost of USD 45 (to replace 50 kg of bone char) per month.

Rural Economic and Educational Development Society / Rural Water Supply and Sanitation Department

REEDS is a voluntary non-profit organization dedicated exclusively to the aim to "Develop, Promote and Implement sound socio-economic strategies to improve the quality of life of Rural Communities". It is currently working closely with the government, encouraging low-cost and easy-to-maintain technologies for safe drinking water.

REEDS is basically a technology provider for water purification. The plant size varies from less than 1 million to more than 10 million litres per day and multiples of this. The society maintain each system for two years under the initial contract and then hands it over to RWSSD for regular operation and maintenance. Water quality monitoring is carried out regularly, and for this, testing kits are provided to the local staff. In the case of small units where only drinking water is treated, water is provided free of cost at the plant.

Economics of the water treatment plants

Of the eight sample villages, where water treatment plants are established, two villages are providing the treated water free of cost. In the remaining six villages water is priced, and plants are run in a business mode. These include an RO plant in Dandu Malkapur, which is maintained by the gram panchayat (village council); an RO plant in Kasaram and UV plants in Bomminampadu and Telaprolu, all maintained by the Naandi Foundation; and an RO plant in Kandlakoya and a UV plant in Pudur, maintained by the Byrraju Foundation.

Measurement of costs and returns

Setting up a water purification plant involves a one-time expenditure, mainly for machinery and plant construction. These costs incurred by the agency are covered by funds raised through contributions from the village panchayat, membership fees and household contributions towards water coupons. The life span of a plant, during which the benefits are expected to flow, is assumed to be 15 years. Benefits and costs are assumed to be constant over this period. The discount rate is assumed to be 13%, which is the actual interest rate paid by the agencies on the bank loans for this purpose.

Two cash flow measures, net present value and benefit–cost ratio, are estimated using these formulae (Perkins, 1994):

$$\text{NPV} = \sum_{t=0}^{n} \frac{(B_t - C_t)}{((1+r)^t}$$

$$\text{BCR} = \frac{\sum_{t=0}^{n} \frac{B_t}{(1+r)^t}}{\sum \frac{C_t}{(1+r)^t}}$$

where:

NPV = net present value
BCR = benefit–cost ratio
B_t = project benefits in time t
C_t = project costs in time t
r = financial or economic discount rate (13%)
n = number of years the project will operate (15)

For benefit–cost analysis, the financial costs and benefits of these plants are considered. The economic viability of the water purification plants is reflected in the cash flow computing, which is often used to make investment decisions (Table 4). Benefit–cost ratios vary across villages and plants. At the present level of coverage, two of the sample villages have favourable net present value and benefit-cost ratios at a 13% discount rate. The benefit–cost ratios indicate that at the present level of coverage, only two of the sample villages are viable. The unfavourable cash flows in the Naandi Foundation plants are mainly due to high capital costs and low coverage levels. Coverage levels have not exceeded 64%, even in Bomminampadu, the oldest UV plant. Though increasing the number of households covered

Table 4. Cash Flow Measures: Benefit–Cost Ratios and Net Present Values, Andhra Pradesh, India.

Village	No. of households	No. of households buying water from the plant	Net present value (INR)		Benefit–cost ratio	
			Without contributions	With contributions	Without contributions	With contributions
Dandu Malkapur	1,032	950	2,025,498	2,112,455	3.37	3.75
Kasaram	261	240	−510,694	−380,258	0.52	0.60
Kadlakoya	507	460	398,973	1,312,016	1.18	2.01
Pudur	900	300	−113,943	175,000	0.92	1.33
Bomminampadu	500	320	−1,584,380	---	0.34	---
Telaprolu	3,000	700	−1,124,175	−672,000	0.59	0.71

Source: Authors calculations based on the data from the respective plants.

in these villages not only makes economic sense but also helps in improving the micro-environment and social welfare, it appears to be hard to do, especially in the coastal districts. Given the low price elasticity of demand for treated water, awareness building has the potential to increase the coverage without losing on pricing (Reddy et al., 2008).

In the majority of cases, part of the capital cost comes from individual donations and household membership fees. The cash flow measures are estimated with and without these contributions. With these contributions there is overall improvement, and Pudur Village has a favourable cash flow. There is no change in the viability status of the Naandi Foundation.

Institutional dynamics

The institutional models for managing treated water provision in the sample villages can be categorized into four kinds (Table 5). The Sai Oral Health Foundation operates the community water purification plants on a small scale and focuses on areas severely affected by fluorides. It also supports the construction of roof water harvesting structures through provision of financial support (subsidy) and arranging bank loans. It works directly with communities and individuals. Though it receives support from the gram panchayat in terms of getting consent to establish the plant and access the water source, the systems are managed entirely by the community workers of the foundation. Each worker is responsible for four or five villages in the neighbourhood. The community worker visits the plants regularly and checks their functioning. These workers are responsible for monitoring the water quality and changing the bone char as necessary. Since the operation and maintenance of the filters are carried out with the foundation's own resources, people are happy about the services.

REEDS established the RO plant in Dandu Malkapur with contributions from the community, a grant from the district collector, and the village panchayat. The plant is fully controlled and managed by the village panchayat and run on commercial basis. The annual system maintenance is given to the TATA projects. One person on a monthly salary is appointed to take care of day-to-day activities, running of the plant, sale of water, etc. Despite being operated by the village panchayat, this plant is among the most efficiently run and has the highest returns. Water charges are the highest in this village (USD 0.044 per 20 litres, against USD 0.03 in other plants). This goes against the general belief that public systems run inefficiently, as people often do not pay for public services. The plant seems to have been working satisfactorily for more than five years, but there is no regular water quality monitoring.

Table 5. Institutional Mechanisms in the Sample Villages in Andhra Pradesh, India.

Village	Total cost (INR)	Community contributions	Role of GP	O&M	Fee collection	Fixing of water charges
Lachammaguda	12,000	None(Sai Oral Health Foundation)	Limited	SOHF	None	None
Dandu Malkapur	500,000	YES (INR 300,000 by community and INR 200,000 by gram panchayat)	100 %	GP (TATA)	GP	GP
Kasaram	700,000	YES (INR 150,000)	Tripartite agreement	NF	NF	NF
Kandlakoya	1,220,000	YES (entire cost: INR 1,220,000; GP provided land)	Limited	GVS (BF)	GVS	GVS
Puduru	700,000	YES (INR 525,000)	Limited	GVS (BF)	GVS	GVS
Bomminanpadu	2,200,000	None (NF)	Tripartite agreement	NF	NF	NF
Telaprolu	2,200,000	YES (INR 440,000, NF, WHI and loan)	Tripartite agreement	NF / WHI	NF	NF
Bhairavapatnam	450,000	None (RWSSD)	100%	GP (REEDS)	None	None

Note: GP = Gram Panchayat; O&M = Operation and Maitenance; SOHF = Sai Oral Health Foundation; GP = Gram Panchayat; NF = Naandi Foundation; GVS = Gram Vikas Samithi; WHI = Water Health India; REEDS = Rural Economic and Educational Society; RWSSD = Rural Water Supply and Sanitation. Source: Authors calculations based on the data from the respective plants.

In the case of Bhiravapatnam, the entire cost of the plant was borne by RWSSD. REEDS maintained the system for the first two years and then handed it over to the village panchayat. Treated water is provided free of cost to all the villagers, who collect the water from the plant in their own vessels and pots. The president of the panchayat is not willing to charge for the water. The workers operating the slow sand filter systems operate and maintain the micro-filters as well. The people in the village were found to be happy with the functioning and effectiveness of the system. Here also there is no regular water quality monitoring. In both the villages where REEDS is working there is no provision for transport of water to households.

The model the Byrraju Foundation adopted is more community-oriented. In both villages, a major share of the project finance was raised through individual contributions (Table 5). Involvement of the gram panchayat was limited to providing space for the plant and giving access to the raw water source for the treatment plant. Initially either the Byrraju Foundation or the technology provider manages plants. The actual capital expenditure of the Byrraju Foundation is low, and it hands the plants over to the village development council once this capital has been recovered. The setting up of the village development council is part of the overall village development activities taken up by the Byrraju Foundation in its adopted villages. In both sample villages, the water treatment plant was handed over to the village development council less than two years from the launch of the project. The village development council takes responsibility for managing the system (with technical support from the foundation). It can fix the water price independently and maintains its own bank account (operated by a treasurer) in which the revenue from the sale of water is deposited. In Pudur, it was reported that the profits from the treatment plant have increased substantially with the increase in the price of water from USD 0.022 to USD 0.030 per 20-litre can. The village

Table 6. Institutional Links in the Byrraju Foundation, Andhra Pradesh, India.

Gram panchayat	Community (village development council)	Byrraju Foundation
• Allotment of land (free of cost) • Permission to draw water from main source • Obtain power connection	• Collection of contributions from community and supervision of construction of building • Identification of youth from village for operation of plant • Oversee operation of plant and distribution of water • Offer creative solutions to increase utilization of water	• Awareness creation, mobilization of resources, develop criteria and identification for eligibility • Identify suppliers of equipment and its procurement and train operators • Process framework for sustenance and maintenance of project; assess, document and disseminate the impact • Quality check and monitoring

Source: Based on the field survey.

development council employs workers to operate and manage the plant and also takes a very active part in its management (Table 6). It was observed that one of the main contributors to the plant is usually the person in charge of the plant under the village development council. Though they do not gain monetarily, they run the plants as their own business. The surplus generated from the plants is expected to be used for the overall development of the village as well as plant maintenance. Both plants are run efficiently and generate positive cash flows. An interesting feature of these plants is that water is provided free of cost to public schools, hospitals, government offices, hostels, etc. Water quality monitoring is done with the help of local colleges as well as through the in-house facilities of the foundation. While microbial tests are conducted at the plant on alternate days, water samples for chemical analysis are sent to the foundation's headquarters. The results of the water tests are displayed for the general public at the plant. All the plants have can cleaning machines, and the water cans are cleaned prior to every filling.

After the plant is stabilized and break-even is reached, its operation is handed over to the council or to a self-help group, under the supervision of the council, within the village. The Byrraju Foundation takes care of quality issues (Table 6). The plant is expected to break even in three to four months at 30% capacity utilization. Water quality monitoring is carried out weekly. Water quality charts for treated water and raw water are maintained at the plant. To maintain quality at the distribution level, water cans are cleaned prior to refilling using chlorine and water jets.

The Naandi Foundation's institutional model is close to a business model, where the contributions of the community are limited. Most of the capital is raised through its own investments or bank loans. It was reported that the foundation also uses public funds, like Member of Parliament or Member of Legislative Assembly funds, as initial capital for setting up the plants. In each village, the Naandi Foundation enters into a tripartite agreement with the gram panchayat and the technology provider, Water Health India (Table 7). The duration of this agreement varies from village to village (five to eight years in the sample villages); some of the presidents were found to be not familiar with or aware of this agreement. Naandi Foundation or Water Health India staff maintain the plants. The foundation is also actively involved in awareness building and promotional activities. Water is sold in 12- and 20-litre cans for USD 0.022 and USD 0.033, respectively. The money is deposited at Naandi Foundation headquarters or collected by Water Health India every week.

Table 7. Institutional Links in the Nandi Foundation, Andhra Pradesh, India.

Community/village panchayat	Technical solution provider	Naandi Foundation
The village panchayat provides a regular source of water and land for the water purification system with the necessary perimeter fence to secure the location of the treatment facility. The panchayat also provides electricity at concessional rates. This arrangement is formalized in a panchayat resolution and a tripartite agreement, which outlines the roles of all the parties involved.	The technical provider leads the construction and installation of water purification plants (at identified village sites). In certain cases it is responsible for hiring personnel to operate and maintain the plants. It continues to provide technical assistance to the plant in the form of an annual maintenance contract.	Naandi acts as the programme manager and coordinator. It ensures that the project meets its deliverables. Naandi is responsible for undertaking campaigns to educate rural communities on the importance of safe drinking water and incentivizing the local community to use the treated water of the community-based safe drinking water system for areas where there is a high level of cross-contamination. This is done through a mobilization drive at the personal, household and community levels, led by the safe water promoter. S/he works with key stakeholders in the village. Upon completion of the construction of the purification plant, Naandi is also responsible for collecting user fees.

Source: Based on the field Survey.

The foundation does not exit the village after installing the plant, as its observation is that providing safe water alone does not ensure healthy lifestyles. Concurrently, a behavioural change campaign is introduced via the safe water promoter for good hygiene practices. Operators are employed by either the technical solution provider or the Naandi Foundation directly to manage the plant on a day-to-day basis.

Despite awareness-building activities, coverage continues to be low in Bomminampadu, even after two years. Initially, more than 500 cans of water were sold per day, but this later fell to 300 cans. One reason for the low coverage is that substantial numbers of households from Scheduled Castes and Scheduled Tribes in this village migrate for long periods for their livelihood. It indicates that the scale of operation is below break-even, as plant size is determined by the number of households. Intensive promotional activities were taken up in the villages to improve the coverage. Water quality (bacteriological, chemical and physical) is monitored regularly. Water samples are sent to private laboratories once a month. There is no third-party evaluation of water quality, though water quality test results are displayed in the plant.

The institutional modalities of the agencies range from philanthropic to pure market-oriented approaches. The philanthropic model adopted by the Sai Oral Health Foundation is used in limited areas, serving a limited number of communities, due to paucity of resources. But it seems to be serving the most deserving communities, in terms of both water quality and the community's economic status. Management is centralized, with the agency controlling all the important matters. The Naandi Foundation also follows a centralized but a professional corporate approach. But plants under such a managerial set-up do not seem to be efficient, due to high capital costs. The sustainability of this model is doubtful, due to poor economic viability and limited community involvement. The community-oriented model of the Byrraju Foundation is not only economically viable due to low capital costs but also sustainable due to the successful takeover of the plant by the village development council of the respective village. The gram panchayat model adopted by RWSSD is equally, if not more, successful in terms of viability and sustainability. This goes against the general belief that gram panchayats cannot manage systems efficiently. This is clearly reflected in

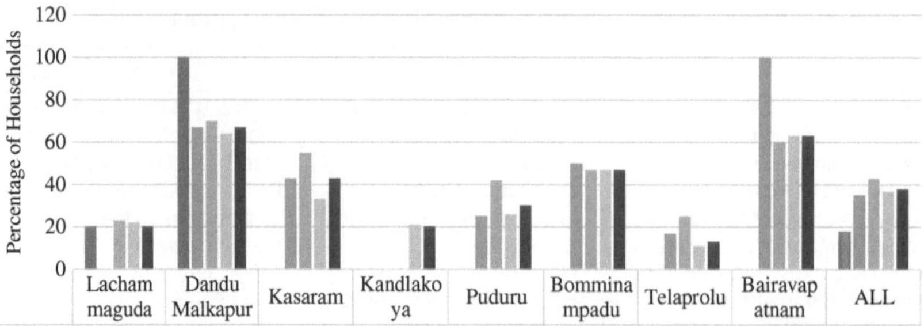

	Lacham maguda	Dandu Malkapur	Kasaram	Kandlako ya	Puduru	Bommina mpadu	Telaprolu	Bairavap atnam	ALL
Large	20	100	0	0	0	0	0	0	18
Medium	0	67	43	0	25	50	17	100	35
Small / Marginal	23	70	55	0	42	47	25	60	43
Landless	22	64	33	21	26	47	11	63	37
Total	20	67	43	20	30	47	13	63	38

Figure 1. People's Perceptions Regarding the Role of Panchayats in Managing Water Treatment Plants, Andhra Pradesh, India.

peoples' perceptions of the capabilities of gram panchayats in managing the plants. In five of the eight sample villages, the majority of people opined that the panchayat is not in a position to manage the water treatment plant (Figure 1). The positive response by more than 60% of the households in two sample villages (Dandu Malkapur and Bhiravpatnam) is borne out of their experience, where the panchayat is running the system successfully.

Provision of treated water in rural areas: public versus private

Drinking water supply is considered a public good and expected to be provided by public agencies. But increasingly, private agencies are entering the domain of drinking water provision and gaining acceptance, even when they resort to selling water, even to the poorest, in spite of its being a social good. What is the rationale behind this? Is this trend in the right direction, or is it desirable? Do private initiatives have anything to offer to enhance the performance of the public agencies? We discuss some of these issues now.

Public provision of drinking water in rural areas is the most common arrangement in all the regions, irrespective of source (surface or groundwater). This includes areas with extensive contamination of groundwater with fluorides and local surface water bodies with bacteriological contamination. But the attempts of the government to provide household defluoridation devices have not succeeded in terms of their spread and acceptance. Similarly, the slow sand filters in the surface systems have proved ineffective due to complicated maintenance. In most villages one finds the filters abandoned, and water from the source pumped directly into the distribution systems. On the whole, the public systems have failed to address the quality issues effectively. This provides fertile ground for the entry of private agencies into the drinking water supply sector. Initially, communities bought bottled water from nearby urban centres to avoid adverse health impacts. The provision of treated water in their villages at low cost has made these private initiatives popular in a number of cases.

These initiatives were indeed effective in reducing the health impacts of untreated water in a number of villages. At the same time, their penetration has been limited to areas with chemically contaminated groundwater. This is mainly due to the serious health problems posed by consuming water with high fluorides. In the case of surface water sources, contamination can be managed through filtering, boiling, etc. Also, there is a possibility of recontamination by pathogens during collection and storage, which does not happen with treated water supplied in cans. People also seem to be averse to paying for drinking water. In one of the cases (Telaprolu), the raw water contamination was not so heavy that it required expensive treatment. This could be a case of profit making rather than people's welfare. Another issue is that these plants provide water only for household drinking and cooking needs, leaving the majority of household water demand unmet, or to be fulfilled by untreated water. Because of this, only waterborne diseases could be tackled at the household level, and water-washed diseases could not be tackled. This is due to the high cost of production of water using these technologies. While there are few alternatives to RO for treating chemically contaminated groundwater, other effective low-cost technologies need to be explored in the case of bacterial contamination.

The public initiatives for the provision of treated water are no less impressive when compared to the private initiatives. The gram panchayat–managed system in Dandu Malkapur Village is the best in terms of economic performance and sustainability. This is the longest-running plant in the lot. In Bhiravpatnam the free provision of treated water using microfilter technology is effective and working well under the management of the gram panchayat. It is hard to conclude that private initiatives are more efficient or effective. Though efficiently run, pricing of treated water is not widely accepted by the panchayats. Under such circumstances, the existence and experience of private agencies in setting up the plants in a business mode helps in creating a demonstration effect. But this model is more appropriate for villages with groundwater sources. For surface water sources, there is need to look at alternative technologies that are cost-effective and functional, like microfilters. Microfilter technologies are capable of serving at any scale to meet the full domestic requirements of rural households (around 70 litres per capita per day) rather than only drinking and cooking needs (5 litres per capita per day).

Managing local water bodies is equally important in reducing the incidence of water contamination. Water bodies such as tanks need to be specifically protected to meet drinking water needs. In groundwater-dependent regions, if tanks are available in the villages, they could serve the purpose of providing drinking water directly during the rainy season, as well as recharging the groundwater, which can be used for drinking purposes during the winter and summer months. Recharge from tanks would help reduce the fluoride concentration of the groundwater. Inclusion of such tank systems in the watershed programme, which is widely adopted in Andhra Pradesh, could mitigate the quality and quantity problems of drinking water to a large extent. Rehabilitation of the traditional tank systems and protecting their catchments would go a long way in improving and sustaining the surface water bodies.

Water quality monitoring is another important area where there is a need for better institutional safeguards. Presently, private agencies prefer to establish plants in villages where the community is receptive to their idea and the gram panchayat cooperates with the private agency to enter into a tripartite agreement. Agencies consider these two aspects critical for their success. In the process, the private initiatives tend to bypass the poorest and most severely affected villages, where these conditions are unlikely to be met. As evident in the case of Telaprolu Village, resources may be diverted to less deserving locations with the neglect of severely affected villages. As of now there is no monitoring of the functioning of the private activities in the villages. There is a need to divert private resources to the needy villages and to see that funds are not invested in areas where the need is not so great at present. Similarly, water quality monitoring is carried out in house in most of the cases. Third-party monitoring needs to be established to ensure high-quality service from the private agencies. Institutionalizing water quality monitoring prior to the establishment of the plants and during the operational stage should be taken up on a high-priority basis at the state level. This monitoring cell should be made responsible for permitting the setting up of plants according to the existing water quality in the villages.

Concluding remarks and policy implications

The preceding analysis, supported by interactions with the various agencies involved in the supply of treated drinking water in rural areas, rural communities and other stakeholders clearly indicates that these agencies provide good-quality water at a price which is affordable to most of the households in the villages. The extent of improvements in the conditions of the communities is directly linked to the nature and degree of water contamination in the villages. Three of the agencies (Byrraju Foundation, Naandi Foundation and REEDS) use sophisticated technologies for surface and groundwater purification, while the others (e.g. Sai Oral Health Foundation), which have concentrated in the groundwater regions, use simple, low-cost technologies. While some agencies provide treated water free of cost, others follow different cost-sharing mechanisms – some with high community share (Byrraju Foundation) and some with low community share (Naandi Foundation). They also adopt different institutional approaches to manage the systems. The systems established by the department are handed over (immediately or after the initial maintenance contract with the technology provider) to the gram panchayat. Excepting the Byrraju Foundation, private agencies do not have a proper protocol for turnover of the plants and exit from the village.

This analysis brings out some of the positive and negative aspects of the institutional models adopted by various agencies. Some of the important issues include that demand for good-quality water is mostly from fluoride-affected villages and not in villages affected by bacteriological contamination, though contamination levels (bacteriological and turbidity) are high in the latter. The business model adopted by some of the agencies does not look viable at first glance, but they are trying to make them viable through lower costs (capital costs, maintenance costs, etc.) and higher revenues (increased coverage). While the public and private agencies have very different exit strategies, the difference in the strategies within the private agencies is linked to cost recovery. The philanthropic service providers seem to target the needy and the most deserving villages and communities; but their business model has serious limitations in terms of scalability.

Not all the models are successful, and the sustainability of some of the models is doubtful. In the unsuccessful cases, it is mainly due to the absence of appropriate institutional set-up at the village level, though in some cases the financial viability is poor due to high capital costs. The present study clearly demonstrates the potential of gram panchayats in managing water treatment plants efficiently, but scaling up could run into problems of financial sustainability due to over-riding political resistance to water pricing. But at the same time, evidence indicates that village panchayats could adopt business models and make them successful through appropriate techno-institutional models. For this there is a need for awareness building at the community level as well as in the village panchayat, along with cross-learning between the experiences of private and public agencies.

The models assessed here can be broadly grouped under private (non-governmental) and public–private initiatives. Of these initiatives, the public–private model is more efficient and sustainable. Scaling up these initiatives in groundwater as well as surface water–dependent villages would require some fine-tuning of institutional arrangements and technologies. To avoid political-economy bottlenecks, the village development council model adopted in the Byrraju Foundation villages could be effective. The council needs to be made an apolitical body to run the systems more efficiently, financially as well as institutionally. While RO technology is well suited for fluoride-affected areas, microfilters appear to be more effective for meeting the entire water needs of the household in areas having surface water sources. In fact, microfilters should replace the existing slow sand filters across the villages.

The following issues and concerns need to be addressed for scaling up of the activities at the macro level.

Formalizing the role of government. Presently, the government does not play an active role. This is resulting in adverse selection of villages by the private foundations. In the process, more deserving villages get sidelined. In this context, the issuing of guidelines by the public authorities for the entry of private agencies into villages is necessary. Private agencies could be encouraged to provide necessary support to the public systems, rather than making it a profitable business to supply treated water. Further, financial incentives could be provided for providing service to the most difficult and deserving villages. The process should start with identifying the most deserving villages and targeting them through some kind of incentives. More importantly, protecting the existing water bodies would substantially address the quantity and quality issues in the rural areas.

Water quality monitoring. Institutionalization of water quality monitoring is necessary to ensure better service delivery. Water treatment plants should be subject to third-party monitoring to check the quality of water supplied. This monitoring should be under the purview of the state nodal agency. Scale and mode of operations, pricing, etc., ought to be disclosed by the agencies before they are issued a no-objection certificate to operate in a village. The water quality monitoring agency also should certify that the specific village is in need of a treatment plant (Reddy & Khurian, 2010).

Public–private partnership. This model could be in the form of build-operate-transfer, whereby private agencies build the plants and operate them till they fully recover their investments, and then hand over the systems to the communities. Initially, communities should be encouraged to share the capital costs so that the handing-over can happen quickly. Alternatively, the agencies could identify social entrepreneurs who can run the plants over the long term. These entrepreneurs could be identified and trained by the agencies. The department should play the role of facilitator as well as monitoring authority.

Role of panchayati raj institutions. Though the panchayati raj bodies are effective in managing the systems, in some cases political-economy factors (like populism) could become bottlenecks with respect to making it financially viable. The model of village development council adopted by the Byrraju Foundation could be used to overcome the political sensitivities of water pricing and maintaining the systems efficiently. However, there is a need for synergy between the panchayati raj and village development councils for better accountability, transparency, etc. While the council could be an implementing agency, the panchayat could act as a monitoring agency at the village level. This could be a part of the institutional arrangement, rather than a separate institutional entity. In the absence of community ownership, these plants become water-selling kiosks. This could become an incentive to private agencies to capture the market for drinking water. Effective involvement of the village panchayat and the line departments in the enterprise would reduce the risk of monopolistic water markets.

Inclusiveness. In the villages supplied by surface water sources, the coverage of households for treated water is 23–66%. The remaining households still depend on poor-quality water. Though there is no clear bias against the poor households, expanding the coverage appears to be difficult in these regions due to perceptions that the benefits of treated water are not worth the cost and effort. Besides, in some of the well-off villages households use individual purification devices (filtering, boiling, etc.). In the case of groundwater sources the coverage is quite high due to the high-risk nature of contamination. Villages with fluoride-contaminated groundwater and no alternative source of water should be given high priority in expanding the coverage through RO plants. The scale of operation in these villages could be limited to drinking water. On the other hand, adoption of microfilter technology to minimize contamination in the case of surface water could replace the present slow sand filters to provide water for all purposes, i.e., above 40 litres per capita per day. Even in the context of inclusiveness, microfilter technology fostered through the public–private institutional model is expected to be more effective for scaled-up operation.

Finally, the policy prescriptions presented here are constrained by the limitations of the study. The coverage of the study is limited to eight villages, and hence it may not represent the different socio-political contexts that are prevalent. The benefit–cost analysis in this study is limited to computing financial costs and revenues and does not include a comprehensive economic analysis incorporating the value of the health benefits of consuming treated water. Future studies need to take these aspects into account in order to provide more insight into the problem.

Acknowledgements

Thanks are due to the two anonymous reviewers, Dr M. Dinesh Kumar, and the editor-in-chief of the *International Journal of Water Resources Development* for valuable comments and suggestions. However, the usual disclaimers apply.

Disclosure statement

No potential conflict of interest was reported by the author.

References

Athena Infonomics. (2012). *Public private partnerships in India: Lessons from experiences*. Chennai: Public Policy Team, Athena Infonomics India Pvt. Ltd.

Gopakumar, G. (2010). Transforming water supply infrastructure regimes in India: Do public-private partnerships have a role to play? *Water Alternatives, 3*, 492–511.

Government of India. (2006). *Guidelines for national rural drinking water quality monitoring and surveillance programme*. New Delhi: Rajiv Gandhi National Drinking Water Mission, Department of Drinking Water Supply, Ministry of Rural development, Government of India.

Lloyd Owen, D. A. (2016). Public–private partnerships in the water reuse sector: A global assessment. *International Journal of Water Resources Development, 32*(4), 1–10. doi:10.1080/07900627.2015.11 37211.

Naiga, R., Penker, M., & Hogl, K. (2015). Challenging pathways to safe water access in rural Uganda: From supply to demand-driven water governance. *International Journal of the Commons, 9*, 237–260. doi:10.18352/ijc.480

Perkins, F. (1994). *Practical cost benefit analysis*. Melbourne: Macmillan Education Australia PTY Ltd.

Pushpangadan, K. (2006). Drinking water and well-being in India: Data envelopment analysis. In V. R. Reddy & S. M. Dev (Eds.), *Managing water resources: Policies, institutions, and technologies* (pp. 179–196). New Delhi: Oxford University Press.

Reddy, V. R., & Kurian, M. (2010). Approaches to economic and environmental valuation of domestic wastewater. In M. Kurian & P. McCarney (Eds.), *Peri-urban water and sanitation services: Policy, planning and method* (pp. 213–242). Netherlands: Springer.

Reddy, V. R., Kullappa, M., & Rao, D. M. (2008). Social cost-benefit analysis of improved water quality in rural areas: an exploratory study in coastal AP. *Journal of Social and Economic Development, 10*(1), 68–97.

Wate, S. R. (2012). An overview of policies impacting water quality and Governance in India. *International Journal of Water Resources Development, 28*, 265–279.

van Wijk-Sijbesma, C., & van Dijk, M. P. (2006). *Water and sanitation: Institutional challenges in India*. New Delhi: Manohar Publishers.

Financial performance of India's irrigation sector: a historical analysis

A. Narayanamoorthy

ABSTRACT

India's public irrigation sector is one of the largest in the world in terms of number of large reservoirs, total storage capacity and irrigated area. But poor financial performance has been threatening its sustainability. Hence, many changes have been introduced in the area of water pricing over the years. But studies that focus on the issue of financial recovery are scanty. Analysis presented in the article shows that despite a substantial increase in area under irrigation, there has been a consistent decline in revenue generated from irrigation fee collection across states. The recovery rate of irrigation fees has been better in less developed states than in more developed states.

Introduction

The major aim of this article is to bring out the real facts about the recovery rate in the irrigation sector in India. Irrigation is critical to agricultural growth in the arid and semi-arid tropics. Irrigation increases land use intensity and cropping intensity and allows farmers to use modern technology, all of which significantly increases land/crop productivity (Kumar, 2010; Vaidyanathan, 2013). Irrigation has also reduced rural poverty in a sustained manner by activating crop cultivation in many Asian countries, including India (Hussain & Hanjra, 2004; Narayanamoorthy, 2001, 2007a; Saleth, Namara, & Samad, 2003). Given the strong link between irrigation and rural poverty reduction, Indian governments have made massive investments in irrigation facilities since the first five-year plan. Up to the end of the 11th plan period (Planning Commission, 2013), public investment in the sector was over INR 4819 billion (USD 1 = INR 68 as of 24 January 2017) at current prices, in which the major and medium irrigation (MMI) sector alone accounted for about 73% (Planning Commission, 2013). This resulted in a substantial increase in the area under irrigation, from 22.56 million ha in 1950–51 to over 92 million ha in 2012–13; the surface irrigated area expanded from 11.92 million ha to about 18 million ha during the same period (for details, see Kumar, Narayanamoorthy, & Singh, 2009; Narayanamoorthy, 2007b).

The impact of irrigation on agricultural development and overall socio-economic conditions in India is profound (Jagadeesan & Kumar, 2015; Vaidyanathan, 2013). However, the

This paper was originally published with errors. This version has been corrected. Please see Corrigendum (http://dx.doi.org/10.1080/07900627.2017.1315990)

financial condition of the sector has been deteriorating over the years, partly due to rising operation and maintenance (O&M) expenditures and partly due to the low water rates, which remained almost stagnant for several decades. The financial rate of recovery, which was close to 100% in 1975–76, fell to about 20% by 2013–14 (CWC, 2015). Time series data suggest that the irrigation sector made a net positive contribution to government finance up to the early 1950s, but the situation had completely changed by 1967–68. The losses accelerated during the 1980s, and expenditures on O&M for MMI projects exceeded the revenues from water charges by 1988–89 (World Bank, 2005). Therefore, the issue of improving financial recovery has caught the attention of policy makers for quite some time now (Gulati, Meinzen-Dick, & Raju, 2005 Ray, 2005; World Bank, 2002).

The public irrigation sector cannot be sustained unless it generates adequate revenue, because of the great expense required for the upkeep of the mammoth infrastructure, which is crucial to their physical performance. Much of the public irrigation infrastructure has been crumbling, owing to paucity of funds for O&M. A recent study by the World Bank (2005, p. 42) underlined:

> The cost of replacement and maintenance of India's stock of water resource and irrigation infrastructure would be about $4 billion a year, which is about twice the annual capital budget in the five year plan.... Not more than a tiny fraction of this is actually spent on asset maintenance and replacement.

Studies focusing on the issue of irrigation cost recovery mostly point to two reasons for the poor recovery rate: higher O&M expenditures; and low and unrevised water rates (Narayanamoorthy, 2007c; World Bank, 2005).

Studies of financial aspects of the irrigation sector, especially in India, have heavily favoured raising water rates to improve cost recovery, though a huge increase in O&M costs is one of the prime reasons for this sorry state of affairs (World Bank, 2002, 2005). The Vaidyanathan Committee report on pricing of irrigation water (GOI, 1992) underlined the fact that the recovery rate has declined drastically less due to low and unrevised water rates and more due to a huge increase in O&M expenditures. It surmises:

> The staff component has been increasing over the years leaving progressively less funds for physical maintenance. The proliferation of personnel has occurred on account of various causes (such as liberal norms of staffing, obligation to provide permanent employment to workers from the construction stage, political influence and intervention from courts). Further, increases in staff costs also result from the rapid rise in wages and emoluments. The combined effect is that in the country as a whole, the staff component has risen from 34 percent in 1974-75 to 43 percent in 1986-87; it is mounting from year to year. This cannot be permitted. Deliberate efforts are called for to bring down the staff costs substantially. (pp. 104–105)

However, a recent estimate on the MMI sector for the period 1992–93 to 2013–14 showed that out of the total working expenses of INR 2,181,443 million, only about 31% was spent on staffing (direction and administration), and the rest on other items, such as O&M, machinery and equipment, and survey and investigation (CWC, 2015).

Deshpande and Narayanamoorthy (2001) also found that despite water rates higher than the national average, the recovery rate in Maharashtra State (this state alone accounts for about one-third of large dams in India) was only 3.66% in 1991–92 because working expense per hectare of irrigation was more than five times the national average. A World Bank (2002) study on irrigation-sector performance in Maharashtra observed that spending on O&M for existing facilities is already inadequate because of financial constraints, but salaries account for 70% of total O&M costs.

Although improved service quality, accountability for expenditures, realistic assessment of irrigation charges, and user participation are very important for increasing the financial recovery of the irrigation sector, studies have not addressed these issues (Gulati et al., 2005; Ray, 2005). Importantly, no detailed studies exist on financial recovery involving national and state-level data. A few states have introduced reforms in the irrigation sector, including pricing reforms. The state of Maharashtra has initiated very bold reforms to bring radical changes in the overall performance of the sector, though no detailed studies are available on the same. Keeping this in view, the present study makes an attempt to address the state of financial recovery in the irrigation sector, with the following objectives: (1) to assess the overall financial performance of the irrigation sector at national and state levels; (2) to analyze the trends in gross receipts and working expenses as well as in recovery rate across different time periods; (3) to carry out an in-depth study of a state that has substantially improved financial recovery in recent years; and (4) to analyze the nature of the relationship between agricultural development (ability to pay) and recovery rate across major states in India.

Data and methodology

The study specifically focuses on financial recovery in the MMI sector of India, which accounted for about 73% of total irrigation investment as of March 2012 (Planning Commission, 2013). Water being a state subject in India, each state follows its own policy regarding pricing of irrigation water and allocation of water across crops and seasons. Besides, the level of irrigation development and the demand for irrigation water are not the same across the Indian states owing to the variation in agro-climatic conditions and cropping patterns (GOI, 1992). In view of this, the analysis covered a total of 12 major states with large areas under canal irrigation, representing four different regions (three states in each region). Since this is a macro-level study, secondary data published by government agencies were used. Data on financial aspects of the irrigation sector are published with a 3–5 year time lag, and therefore, data from 1974–75 to 2013–14 were used. (Time series data used in the analysis are presented in the Appendix.) Data relating to water prices, investment in the irrigation sector, O&M charges, receipts from irrigation services, and others were collected from various publications of the Central Water Commission (CWC, http://www.cwc.gov.in). Statistics pertaining to agriculture and irrigation development were obtained mainly from publications such as *Agricultural Statistics at a Glance* and *Indian Agricultural Statistics,* both published by the Ministry of Agriculture, Government of India, New Delhi (http://www.agricoop.nic.in).

To estimate financial recovery in the irrigation sector, we followed the methodology adopted by the CWC, which considers gross receipts against expenditures for O&M works. To study the trends in O&M costs as well as gross receipts from the MMI sector, growth rate (log-linear) was computed. Since the focus of the study is to find out the trends in cost recovery, all the analysis was carried out for current prices. To examine the nature of the relationship between agricultural development (used as a proxy variable to reflect farmers' ability to pay) and financial recovery across states, the financial recovery rate of different states was compared against the canal-irrigated area as well as against the value of crop output (INR/ha) at different time points. Simple correlation analysis was carried out to characterize the association between financial recovery and stage of agricultural development.

Financial recovery at the national level

India is a vast country with a high degree of heterogeneity in terms of socio-economic conditions across regions. Moreover, each state follows its own irrigation policy, including the norms for charging for irrigation water. Therefore, the financial performance of individual states is likely to be different from the aggregate situation at the national level. In view of this, we first analyzed the financial performance of the irrigation sector at the country level to understand the overall scenario, which was followed by state-level analysis. Here, taking data covering the period from 1974–75 to 2013–14, the recovery rate was worked out as mentioned in the methodology section. The results of the analysis are presented in Table 1, which shows that the gross receipts from the MMI sector increased by 45 times over the years, from about INR 840 million in the triennium ending (TE) 1976–77 to INR 37,860 million in TE 2013–14. Over the same time, working expenditures increased 204 times, from about INR 1010 million in TE 1976–77 to INR 206,410 million in TE 2013–14. As a result of this rapid growth in working expenditures, the loss incurred by the MMI sector increased substantially over the years. The estimated loss, excluding interest on capital outlay, was only about INR 160 million during TE 1976–77, but increased to INR 168,560 million during TE 2013–14. The growth in working expenditures was even faster than the growth in canal-irrigated area, which increased only from 13.52 million ha in 1974–75 to 16.28 million ha in 2013–14.

The growth in working expenditures, much faster than in gross receipts, meant deteriorating financial recovery over the years. Although many states revised their water rates during this period (for details, see CWC, 2010), this had no impact on the overall situation at the national level with respect to financial recovery rate. However, for the first time since 1984–85, the financial recovery rate did register an improvement, to about 20%, in 2013–14. A relatively smaller increase in working expenditures in relation to the increase in gross receipts during the period helped improve the situation.

The irrigation policies followed during the different time periods considered in the analysis were not the same in India. Till the early 1990s, farmer participation in irrigation management was not visible. But this has changed since the 1990s, when its importance has increasingly been recognized due to pressure from various quarters (for a detailed review, see Vermillion, 1997; IWMI, 2011). In states such as Maharashtra, Gujarat and Andhra Pradesh, tertiary-level management of a significant portion of the canal-irrigated area has already been transferred from the state agency to the water user associations (WUAs). An attempt was made to see whether any perceptible changes in financial recovery had occurred as result of this policy

Table 1. Trends in gross receipts and recovery rate for major and medium irrigation projects in India (INR millions at current prices)

Year	Gross receipts (GR)	Working expenses (WE)	Profit (GR – WE)	Recovery in % (GR/WE × 100)
TE 1976–77	840	1,010	−168	82.62
TE 1980–81	1,040	1,740	−697	62.40
TE 1984–85	1,370	2,820	−1,446	49.50
TE 1990–91	2,010	22,760	−20,751	8.83
TE 1994–95	4,140	36,990	−32,855	11.19
TE 2000–01	5,550	79,860	−74,305	6.95
TE 2006–07	13,210	82,800	−69,584	16.10
TE 2013–14	37,860	206,407	−168,548	18.40

Note: TE = triennium ending.
Sources: CWC (2005, 2007, 2010, 2015).

Table 2. Growth rate in financial recovery for the major and medium irrigation sector during different time periods in India.

Period	Cumulative capital expenditures	Gross receipts	Working expenditures	Recovery rate
1974–75 to 1984–85	14.4	6.4	13.3	−6.8
1984–85 to 1994–95	10.4	11.3	25.4	−14.1
1994–95 to 2004–05	9.8	10.4	5.1	5.3
2004–05 to 2013–14	13.0	14.3	13.1	1.2
1974–75 to 2013–14	11.5	10.2	14.8	−4.6

Notes: Growth rate is computed using log-linear function ($\ln(Y) = a + bt$); all values are significant at the 1% level.
Sources: Computed using CWC (2007, 2010).

reform. For this, we computed the growth rate (log-linear) of variables related to financial recovery for five different time periods: 1974–75 to 1984–85; 1984–85 to 1994–95; 1994–95 to 2004–05; 2004–05 to 2013–14; and 1974–75 to 2013–14.

As expected, the growth rates for the two parameters are not the same in different periods (Table 2). During 1974–75 to 1984–85, the annual growth rate of gross receipts (6.4%) was much smaller than that of working expenditures (13.3%). Therefore, the financial recovery in percentage terms registered a negative growth of –6.8%/y. The situation worsened during later periods. Gross receipts registered an annual growth of 11.3% during 1984–85 to 1994–95, but annual working expenditure registered a growth of 25.4%. This high growth rate in O&M expenditure had a deleterious impact on the recovery rate, which declined at 14.10%/y during 1984–85 to 1994–95.

Among the different periods considered for the analysis, the growth rate for the most recent periods (i.e. from 1994–95 to 2004–05) shows a very encouraging picture. Gross receipts registered an unprecedented annual growth of 10.4%, and working expenditures recorded a decrease of 5.1%. The sharp deceleration in the growth rate of working expenditures and the very high growth in gross receipts helped to register a 5.3% growth in financial recovery during 1994–95 to 2004–05. Though there are variations in the growth rate of different periods, the overall growth rate computed for the entire time from 1974–75 to 2013–14 shows a very dismal picture, with recovery rate registering a negative growth of 4.6%/y. It appears that the faster growth in working expenditures as compared to gross receipts is adversely affecting the financial performance of the MMI sector. More detailed disaggregated studies are essential to find out why working expenditures in the irrigation sector are increasing so rapidly in India.

Financial recovery at the state level

The area irrigated by MMI schemes accounts for a major share of the surface irrigated area in almost all the states in India, though the proportion varies substantially between states. To study the financial performance of the irrigation sector in individual states, we considered 12 major canal-irrigated states, three from each of the four regions (east, north, south and west). Unlike the national-level analysis, data on various parameters pertaining to financial recovery are not available continuously from 1974–75 to 2013–14 for all the states considered for the analysis. Therefore, the state-level analysis was carried out with data for the period 1987–88 to 2013–14 to find the recovery rate and its changes in different periods. Though our objective here is to study the performance of individual states, an attempt is also made to examine whether any perceptible pattern is emerging across the four regions.

Table 3. Financial recovery of major and medium irrigation projects for selected major states in India.

Region	State	Period (average of two years)	Gross receipts (INR millions)	Working expenses (INR millions)	Recovery (%)
West	1. Bihar	1988–90	63.00	582.50	10.50
		2012–14	183.00	3,924.00	4.70
	2. Orissa	1988–90	51.00	147.50	35.00
		2012–14	4,119.50	5,069.00	81.30
	3. West Bengal	1988–90	15.50	435.00	3.60
		2012–14	95.00	2,584.00	3.70
North	4. Haryana	1988–90	145.50	1,050.00	13.86
		2012–14	1,171.00	9,564.00	12.20
	5. Punjab	1988–90	170.00	717.50	23.69
		2012–14	584.50	9,473.50	6.20
	6. Uttar Pradesh	1988–90	335.00	3,275.00	10.23
		2012–14	3,355.00	29,860.00	11.20
South	7. Andhra Pradesh	1988–90	203.00	4,032.50	5.03
		2012–14	2,000.00	83,822.00	2.39
	8. Karnataka	1988–90	152.00	1,460.00	10.41
		2012–14	307.00	3,371.00	9.10
	9. Tamil Nadu	1988–90	14.50	724.50	2.00
		2012–14	322.20	7,852.80	4.10
East	10. Gujarat	1988–90	169.50	3,269.00	6.00
		2012–14	8,058.00	5,427.00	148.50
	11. Madhya Pradesh	1988–90	160.00	570.00	28.07
		2012–14	1,381.00	6,410.00	21.55
	12. Maharashtra	1988–90	231.50	3,993.50	5.80
		2012–14	5,105.00	8,025.00	63.61

Sources: CWC (2005, 2007, 2010, 2015).

The three major states chosen from the eastern region are Bihar, Orissa and West Bengal. These three states together accounted for about 19% of the canal-irrigated area in the country in 2013–14. Table 3 presents the details of financial results for two time periods: the late 1980s (average of two years, 1988–89 and 1989–90) and the mid-2010s (average of two years, 2012–13 and 2013–14). The analysis shows that financial performance varies widely amongst the three states. While financial recovery is generally poor for all the states in both time periods, it is relatively better in Orissa, which is one of the agriculturally less developed states. The state also has high rural poverty. Interestingly, the recovery rate here is also higher than the national average. Among the three states, West Bengal is the poorest performer (lower than the national average at both time points), hovering only around 3.56–3.67%.

The three major states from the northern region are Haryana, Punjab and Uttar Pradesh. Historically, canal irrigation has been prominent in these states; together they accounted for about 30% of India's total canal-irrigated area in 2013–14. Given the extensive and intensive use of canal water for irrigation and the higher agricultural productivity, intuitively, financial recovery should be better than in the eastern states. But the results do not show any such robust trend. The recovery rate was very low in all three states in all three time periods. While the recovery rate of Uttar Pradesh marginally increased between the late 1980s and the mid-2010s, it declined from 23.7% in the late 1980s to 6.2% in the mid-2010s in Punjab. A similar trend is observed in Haryana, which is another rich state in terms of agriculture. A higher rate of growth in working expenditures than in gross receipts reduced the recovery rate in all three states.

The southern states, Andhra Pradesh, Karnataka and Tamil Nadu, also show poor recovery rates. Though minor irrigation tanks have historically been the major source of irrigation in the southern region, canals accounted for over 20% of the total canal-irrigated area of the country in 2013–14. In the case of Andhra Pradesh the recovery rate has been less than 3% since the late 1990s. The maximum recovery rate achieved by Andhra Pradesh during the entire period was only about 8%. Similarly, the maximum recovery rate achieved by Tamil Nadu was only about 4%. Among the three southern states, the financial recovery status of Karnataka is relatively better, where the recovery rate hovered around 10.4% in the late 1980s and 9.1% in the mid-2010s. This is because Karnataka is able to control working expenditures, a feat the other two states could not achieve. Separate analysis needs to be carried out to find out why the southern states have performed so poorly in terms of recovery rate. We also need to find out how Karnataka reduced its working expenditure in recent years. The relatively better recovery achieved in Karnataka again reinforces our argument that controlling working expenditures for irrigation projects is essential for better financial recovery.

The three western states, Gujarat, Madhya Pradesh and Maharashtra, accounted for about 18% of total canal-irrigated area of the country in 2013–14. The vast majority of the MMI projects of the country are located in these three states. Capital expenditures for the development of the MMI sector are also large in these three states. Unlike the other three regions, the financial results for western Indian states show a very encouraging picture. In the mid-2010s, Gujarat was able to achieve over a 148% recovery rate, while Maharashtra recorded a financial recovery of over 63.6% (it had about 106% in the late 2000s). This unprecedented achievement in financial recovery was not made by any other states analyzed by us since the late 1980s. The recovery rate for Madhya Pradesh, the lowest-performing of the three, was also higher than the all-India average of 18.4% for 2012–14. What is interesting here is that the performance of Gujarat and Maharashtra was lower than the all-India average till the late 1990s. For instance, in Gujarat, the recovery rate was limping along at 6–8% between the late 1980s and the late 1990s, and it was around 3–5% in Maharashtra during the same period. This was much lower than the rate in the 2000s. For instance, in the case of Gujarat State, the recovery rate increased from 7.9% in 2000–01 to 117% in 2009–10, and it increased from 40% to 113% during this period in Maharashtra State. To understand the reasons for this impressive performance, an in-depth study of Maharashtra State was carried out and is presented in the next section.

Improved recovery in Maharashtra State

Maharashtra is one of the Indian states that has introduced various reforms to improve the overall performance of the irrigation sector since 2000–01 (Table 4). For instance, the Water Regulatory Authority was established to bring transparency to the functioning of the irrigation sector, and water auditing was introduced to improve the efficiency of water use. Water pricing has also been revised on a regular basis to increase the revenue of the sector. In this section, we examine whether the reforms introduced in the irrigation sector helped in any way to improve the financial recovery situation. Though quite a few states have initiated some reforms in irrigation sector, Maharashtra was purposively chosen for studying the impact of irrigation reforms for two reasons. First, the data required for the analysis were readily available for Maharashtra. Second, since Maharashtra's irrigation sector is the largest

Table 4. Summary of reforms initiated in the water sector in Maharashtra State since 2001.

	Major reforms	Specific reform activities
1	Policy reforms, legal reforms and institutional reforms (2001 and 2005)	State water policy formulated with focus on IWRM Maharashtra Management of Irrigation System by Farmers Act of 2005 Water Regulatory Authority, 2005
2	Administrative reforms (2002)	Water auditing introduced Water prices revised to 2–2.5 times earlier rates Manual for operating and maintenance of irrigation systems released On-farm development manual released Project monitoring cell established
3	Application of state-of-the-art technology (2002)	Irrigation status report prepared and published annually Benchmarking report published annually Promotion of water-conservation technologies Use of information technology Research and development to provide solutions to problems faced by field personnel and farmers
4	Capacity building and people awareness campaign (2002)	Capacity building of irrigation personnel Study group (quality circles) Workshop and conferences Capacity building of farmers Water user association formation Involvement of non-governmental organizations
5	Canal Cleaning Movement (2002)	This campaign enabled best use of machinery for productive work. Sugar factories in the command area also participated in the movement. The massive campaign, involving users, NGOs and people's representatives, resulted in *shramdaan* (voluntary physical work) on 30 major, 27 medium and 193 minor irrigation projects in various parts of Maharashtra State. The important aspect of *shramdaan* was that users developed a sense of belonging to the system.

Source: Sodal (2007); Narayanamoorthy (2007c); http://www.irrigation.maharashra.gov.in.

in India, in terms of both the number of large reservoirs and the total live storage capacity (for details, see GOI, 2007), the results can also be useful for the rest of India.

Water rates have been historically very high in Maharashtra when compared to any other major state in India, such as Andhra Pradesh, Tamil Nadu, Punjab, Haryana or West Bengal (CWC, 2010). But the percentage of cost recovery was not appreciably higher when compared to other comparable states till 1999–2000. According to CWC data, the recovery in MMI projects declined from 166% in 1974–75 to 3.9% in 1999–2000 in Maharashtra, while it declined from 64% to 5.7% at the national level during the same period. In fact, in 1999–2000, the recovery rate was 12–20% in Madhya Pradesh, Orissa and Bihar, where the water rates are much lower than that of Maharashtra. An important reason for the poor recovery rate is that working expenses per unit of irrigated area in Maharashtra are about five times the all-India average. For instance, during 1991–92, working expenses were estimated as INR 5627 per hectare for Maharashtra state, but only INR 1032 per hectare for the whole of India (Deshpande & Narayanamoorthy, 2001). However, the policy makers in the state seem to have believed all along that the recovery rate could be increased merely by raising water rates, without the institutional and other reforms required to achieve a higher rate of recovery of water charges from users.

During this period the irrigation sector started facing many serious problems, such as low utilization of created potential, low water-use efficiency, and slow growth in coverage area. All these problems posed a serious threat to the sustainability of the largest irrigation sector (Sodal, 2007). In view of the importance of the irrigation sector in the overall growth of the state's economy, a series of reforms were introduced by the state, covering such aspects as policy, law, institutional development and administration (for details, see Narayanamoorthy, 2007c). These reform initiatives started changing the financial position of the irrigation sector completely in 1998–99, as the recovery rate started increasing considerably (Table 4). According to data from the Water Resources Department, Government of Maharashtra, cost recovery increased from 30% in 1998–99 to over 119% in 2004–05 – but it declined to 27% in 2013–14 according to CWC data.

Though studies in the past have suggested that a good recovery rate can be achieved primarily by increasing/revising the water rates regularly (World Bank, 2002; Gulati et al., 2005; Ray, 2005), this did not work in Maharashtra State. Hence, the state government has introduced a series of policy measures along with the upward revision of water rates since 2000. First, the state revised the water rates, effective September 2001, for both irrigation and non-irrigation purposes in order to fully recover O&M costs. These charges were increased by 15% every year to cover part of the capital cost and also to neutralize the effects of inflation (World Bank, 2002). Second, various measures were initiated to improve the overall supervision of the irrigation system to increase the rate of recovery of water charges since 2001. Targets for every irrigation cycle are fixed at the start of the financial year, and review of recovery is taken in every bimonthly meeting of the Superintending Engineers with the Secretary (Command Area Development). A special drive was also taken up seriously to recover the arrears from non-irrigation users every year (GOM, 2006). Third, since reduction of O&M expenditures was essential to improve financial recovery, efforts were made to minimize O&M costs. As a result, total O&M costs declined by INR 566 million between 1999–2000 and 2004–05 (from INR 4326 million to INR 3760 million), despite a considerable increase in the area under surface irrigation. More importantly, the gap between assessment and recovery of revenue has also narrowed substantially over the years, thanks to the introduction of the project monitoring cell – from INR 2422 million in 2000–01 to INR 50 million in 2005–06 (Table 5). All these helped increase the recovery rate to over 100% since 2002–03. However, according to recent data, the recovery rate of Maharashtra fell to about 61% in 2013–14. It should be noted here that there are discrepancies between the data published by CWC and the government of Maharashtra on cost recovery in the irrigation sector, especially in recent years.

The formation of a large number of WUAs in the state also led to a turnaround in financial recovery. The government of Maharashtra made a policy decision to hand over the management of irrigation systems to WUAs in July 2001. This policy said that water would be

Table 5. Trends in irrigation assessment, O&M cost and recovery in Maharashtra (INR millions).

Year	Assessment	O&M cost	Recovery	Recovery %
1998–99	1,951	3,790	1,135	30
1999–00	2,762	4,326	1,729	40
2000–01	4,375	4,900	1,953	40
2001–02	4,535	4,500	2,516	56
2002–03	4,439	3,700	3,770	102
2003–04	4,533	3,330	3,780	114
2004–05	4,970	3,760	4,480	119
2005–06	4,180	4,530	4,130	91
2010–11	7,667	7,450	7,459	100
2013–14	6,071	8,430	5,150	61

Note: O&M cost includes establishment plus maintenance and repairs.
Source: GOM (2006).

Table 6. Water availability, actual irrigation and water use efficiency in Maharashtra.

Year	Designed water storage (Mm³)	Water available on 15 October (Mm³)	Storage used (%)	Water used for irrigation (Mm³)	Irrigated area on canal (mha)*	Water use efficiency (ha/Mm³)
1999–00	26,716	22,715	85	16,037	1.286	80
2000–01	26,748	18,947	71	13,575	1.298	96
2001–02	28,062	17,817	63	12,346	1.250	10
2002–03	28,715	18,936	66	12,965	1.318	102
2003–04	28,840	16,941	59	10,569	1.244	118
2004–05	28,889	18,298	63	10,603	1.257	119
2005–06	29,110	24,860	85	13,689	1.617	118

* Excluding groundwater-irrigated area. mha = million hectares
Source: GOM (2006); Sodal (2007).

supplied only to WUAs on a volumetric basis; no individual farmers would be supplied water through canals. To provide legal recognition to WUAs, the state enacted the Maharashtra Management of Irrigation System by Farmers Act of 2005 (GOM, 2005). Since the implementation of this act, there has been significant progress in the number of functioning as well as registered WUAs since September 1996. While the number of fully functioning WUAs increased from 100 in September 1996 to 921 in 2005–06, the area operated by these WUAs increased from about 44,000 ha to about 287,000 ha. Thus, the area covered by the fully functioning WUAs increased nearly seven times. Besides using water more efficiently, WUAs also helped reduce the O&M expenditure of the system thanks to free labour from farmers (Bardhan, 2000; Bassi & Kumar, 2011). Importantly, since the WUAs are required to pay the water charges at the stipulated time to ensure water allocation for the next season, the farmers must have paid their required water charges in time, improving revenue collection. The improved quality of irrigation service through the WUAs also ensured farmers' willingness to pay the higher water charges fixed by the water resources department.

Along with higher water rates and other reform measures, water auditing was also introduced. The main objective of the water auditing was to have a proper accounting of water use in various sectors to reduce unaccounted-for water and to increase the revenue of the government. Water auditing on all major projects in the state was made mandatory, as underlined in the Maharashtra State Water Policy of 2003 (GOM, 2003). Water auditing helps compare actual water use efficiency against planned water use efficiency (in ha/Mm³), as well as the losses of water from each system. Since water use has to be accounted for in each system under water auditing, there was tremendous pressure on the system managers to report the area under irrigation precisely, which had not been done earlier. In fact, the initiation of water auditing in the state substantially increased water use efficiency in canal irrigation, from 96 ha/Mm³ in 2000–01 to 118 ha/Mm³ in 2005–06 (Table 6). The increase in area under irrigation must also have increased gross receipts. All these clearly suggest that Maharashtra State was able to achieve a huge increase in financial recovery through all-round reforms in the irrigation sector.

Ability to pay and financial recovery

As we have seen, the financial recovery rate is very low in some states, while it is relatively high in some others. Since farmers' ability to pay for irrigation water is higher in the agriculturally developed states, it was expected that financial recovery would be better in those

Table 7. Canal-irrigated area and financial recovery by states.

State	1974–75 CIA (thousand ha)	1974–75 IFR (%)	1984–85 CIA (thousand ha)	1984–85 IFR (%)	1994–95 CIA (thousand ha)	1994–95 IFR (%)	2013–14 CIA (thousand ha)	2013–14 IFR (%)
Andhra Pradesh	1,590	15.50	1,794	36.90	1,606	17.00	1,429	2.50
Bihar	887	72.50	1,005	29.00	984	22.00	914	5.10
Gujarat	197	71.10	434	25.80	593	7.00	771	148.50
Haryana	1,031	55.30	1,203	46.80	1,382	4.00	1,210	9.60
Karnataka	482	97.60	705	7.00	927	4.00	1,253	9.10
Madhya Pradesh	679	129.30	1,267	42.00	1,825	29.00	1,625	21.55
Maharashtra	339	166.00	794	48.90	499	8.00	1,081	27.60
Orissa	606	30.70	866	134.90	949	12.00	1,245	81.30
Punjab	1,410	63.20	1,399	48.30	1,534	32.00	1,122	6.20
Rajasthan	881	49.60	1,058	19.30	1,427	7.00	1,859	5.46
Tamil Nadu	887	22.50	897	6.80	844	3.00	683	4.10
Uttar Pradesh	2,624	123.50	3,331	169.80	3,142	12.00	2,557	11.20
West Bengal	960	38.20	724	7.30	717	4.00	728	3.70
INDIA	13,484	64.20	16,354	46.00	17,233	10.20	16,278	19.80

Notes: CIA = canal-irrigated area; IFR = irrigation financial recovery.
Source: GOI (various years); CWC (2005, 2007, 2010, 2015).

Table 8. Value of output and irrigation financial recovery by states.

State	1970–73 VOP (INR/ha)	1970–73 IFR (%)	1980–83 VOP (INR/ha)	1980–83 IFR (%)	1990–93 VOP (INR/ha)	1990–93 IFR (%)	2013–14 VOP* (INR/ha)	2013–14 IFR (%)
Andhra Pradesh	4,363	15.50	6,276	6.80	8,728	11.10	11,537	2.50
Bihar	4,010	72.50	4,049	50.67	5,278	20.67	5,670	5.10
Gujarat	4,327	71.10	5,693	27.70	6,640	6.00	11,836	148.50
Haryana	5,090	55.30	6,229	55.13	9,682	12.57	11,569	9.60
Karnataka	4,267	97.60	4,990	39.33	6,342	10.67	6,994	9.10
Madhya Pradesh	2,836	129.30	3,070	38.67	4,406	22.00	5,640	21.55
Maharashtra	2,344	166.00	3,795	88.87	4,490	4.97	5,960	27.60
Orissa	4,073	30.70	4,375	65.37	5,740	28.67	6,690	81.30
Punjab	7,476	63.20	9,708	56.57	13,215	15.50	15,373	6.20
Rajasthan	2,217	49.60	2,335	51.20	3,809	10.67	5,095	5.46
Tamil Nadu	7,900	22.50	8,756	9.70	13,037	2.73	13,117	4.10
Uttar Pradesh	4,590	123.50	5,805	132.93	8,355	8.73	9,894	11.20
West Bengal	5,615	38.20	5,944	11.87	9,507	3.33	12,142	3.70
INDIA	4,257	64.20	5,090	46.80	6,957	9.07	8,460	19.80

Notes: VOP = value of agricultural output; IFR = irrigation financial recovery.
*For the year 2006–07.
Sources: VOP from Bhalla and Singh (2009); IFR from CWC (2005, 2007, 2010, 2015).

states. We examined whether any such relationship exists between agricultural development and financial recovery, using data on canal-irrigated area and recovery rates from the major states (Table 7). Though the agricultural development of a state can be gauged by various factors, we considered two: value of agricultural output (INR/ha) and canal-irrigated area (CIA). For this analysis, data from different time points (1970–73, 1980–83, 1990–93 and 2013–14) from 13 major states that have considerable area under canal irrigation were used, as reported in Table 8.

The relationship between CIA and financial recovery in the irrigation sector across states was first studied. It is evident from the results presented in Table 7 that there is no firm relationship between CIA and irrigation financial recovery (IFR) in the irrigation sector across all four time points: 1974–75, 1984–85, 1994–95 and 2013–14. Since increased availability of canal irrigation would help increase the cropped area as well as crop output per hectare, a strong and positive relationship between CIA and IFR across the states was expected. But the analysis did not show this. In 1974–75, the estimated correlation between CIA and IFR was –0.061, a weak and negative relationship. This situation improved in 1984–85, when the correlation was not only positive but also significant (0.692). This is because most states which had higher CIA also registered higher IFR, except for Orissa. A positive but slightly weaker association between CIA and IFR was also observed in 1994–95. However, the same positive relationship could not be observed across the states in 2013–14. The negative correlation between CIA and IFR suggests that the increased CIA was not accompanied by higher financial recovery.

An indicator of agricultural development which is considered to be related to financial recovery is the value of agricultural output per hectare (VOP). As reported in Table 8, data from 13 major states covering four time points (1970–73, 1980–83, 1990–93 and 2013–14) were used to study this relationship. However, the results do not show any strong positive relationship between VOP and IFR. Most states with higher VOP have consistently low IFR in almost all the time periods. One of the recommendations of the Second Irrigation Commission of India (GOI, 1972) was that the water rate should be based on the value of crop output, suggesting that VOP would be indicative of willingness to pay for water. Therefore, a positive and strong relationship was expected between VOP and IFR. While the coefficient of correlation between VOP and IFR was –0.498 in 1974–75, it was a weak 0.021 in 2013–14. Unexpectedly, the sign of the correlation between VOP and IFR turned out to be negative in all four time points taken for the analysis. This suggests that the states with higher VOP in agriculture (with higher ability to pay) will not necessarily have better financial recovery. There could be many reasons for this. Perhaps in these states farmers have access to alternative sources of irrigation, or perhaps canal irrigation is not very crucial to their agricultural economy.

The data available to us suggest the latter possibility. In Haryana, Punjab, Tamil Nadu, West Bengal and Andhra Pradesh, though VOP per ha of cropped area is quite large, these states also have large areas under well irrigation, including within the canal command areas. It is understood that well irrigation in canal commands is largely sustained by recharge from irrigated fields fed by gravity and canal seepage and that farmers practice conjunctive use in canal commands (Foster, et al., 2010; Kumar, Scott, & Singh, 2011). Import of canal water to these regions helped improve the groundwater condition there, so farmers are less dependent on canal water to sustain irrigated crop production, at least in the short term. In states such as Andhra Pradesh and Tamil Nadu, the farmers in canal commands also have access to water from tanks. In such situations, it will be difficult to raise the water charge, increase the collection rate and improve overall financial recovery, unless institutional reforms are initiated to improve the quality of irrigation service, increase farmers' willingness to pay and reduce O&M charges.

Conclusion and recommendations

The study clearly shows that the financial recovery of irrigation projects has been declining since the mid-1970s at the all-India level, although some moderate improvement has taken place, especially in recent years. The analysis shows that financial recovery is generally low across major states as well. The agriculturally less developed Madhya Pradesh and Orissa appear to have performed better than the agriculturally prosperous Punjab, Haryana and Tamil Nadu. Among the major states, the recovery rate was consistently very low in Andhra Pradesh, Tamil Nadu and West Bengal throughout the period of analysis. A significant improvement in financial recovery was noted in the western states of Maharashtra and Gujarat since 1999–2000. The analysis shows no relationship between ability to pay and recovery rate. The correlation between value of agricultural output per hectare and percentage of financial recovery turned out to be negative in all four time periods considered for the analysis.

There is no doubt that the under-pricing of water severely limits the availability of financial resources for the management of irrigation systems. The performance of irrigation systems cannot be sustained unless they generate adequate revenue to take care of their O&M. Much of the irrigation infrastructure has been crumbling owing to paucity of funds for O&M. The experience of Maharashtra and Gujarat, which turned around the financial recovery situation very recently, shows that the financial working of the irrigation sector cannot be improved merely by revising water rates, but it can be done by institutional reforms that are packaged and sequenced appropriately, along with upward revision of water rates. The ultimate aim of such reforms should be to improve the quality of irrigation service, without which it would be difficult to increase the financial recovery.

The experience of the states which did achieve remarkable recovery in recent years also suggests that to improve the recovery rate working expenditures need to be controlled. For this, cost-cutting measures need to be strictly followed by rationalizing the staff strength per hectare of canal command area. Maharashtra's experience suggests that full-fledged volumetric pricing is possible only after transferring the functions of management and distribution of water to the WUAs. Therefore, concerted efforts are needed to establish WUAs, which can take control of irrigation management below the outlets in the canal system and recover water charges from farmers.

Water accounting at the project level would provide a detailed account of water use in various sectors and actual information about conveyance losses in each irrigation scheme, including unaccounted-for losses or water theft. Water auditing creates an incentive among the officials of the irrigation bureaucracy to enhance their performance to expand irrigation coverage and reduce losses from the system. By increasing water use efficiency and irrigated area substantially, water auditing helps increase financial recovery as well. Except for Maharashtra (GOM, 2006, 2006a), no state seems to have seriously implemented a water auditing programme. The cropping pattern followed in a state is also expected to influence the recovery rate of the irrigation sector; this may have happened in Maharashtra State. Studies need to be carried out to see the influence of cropping pattern in the recovery rate of the irrigation sector.

There is also a wide gap between demand raised and actual collection of irrigation receipts in different states, which is also one of the reasons for the poor cost recovery. Maharashtra State was able to turn around cost recovery mainly by improving the collection rate of irrigation-water charges. Since poor collection of irrigation charges is mainly due to administrative tangles, a serious effort is needed to improve the functioning of administrative machinery, with clear-cut collection targets. Policy makers must understand that no single reform in the irrigation sector will improve the financial condition of the sector. Efforts to control working expenditures should be accompanied by legal and policy reforms, and by institutional development in the sector aimed at increasing the willingness to pay for water and improving the (water charge) recovery rates so as to make a turnaround in financial recovery possible.

Finally, and more importantly, there are serious limitations in conducting performance analysis of the irrigation sector purely on financial grounds. The issue has to be looked at from the farmer's perspective as well. With fluctuating output prices and rising input prices, one has to relate the irrigation charges to farmers' net income as well. It is here that crop pattern has a relationship with financial performance, as cash crops generate higher irrigation revenue as compared to food-grain crops such as rice and wheat. While macro-analysis can conceal the influence of cash crops on financial recovery rate, their effects are fundamental. Another contextual factor that is equally important is the role of hydrological linkages between canal water supply and the recharge of groundwater aquifers. How one can consider the issue of irrigation charges under this condition is still an unaddressed question in the literature. Therefore, to find out the real performance of the irrigation sector, one should also consider all the direct and indirect benefits of the irrigation sector along with financial recovery.

Acknowledgements

This article is a fully modified and updated version of a larger study, "Financial Performance of India's Irrigation Sector: A Macro Level Analysis", sponsored by the International Water Management Institute (Colombo, Sri Lanka) through the IWMI-TATA Water Programme. The author wishes to thank two anonymous referees, Dr M. Dinesh Kumar and the editor-in-chief of the journal, Dr Cecilia Tortajada, for their constructive comments on earlier versions of the article.

Disclosure statement

No potential conflict of interest was reported by the author.

References

Bardhan, P. (2000). Irrigation and cooperation: An empirical analysis of 48 irrigation communities in South India. *Economic Development and Cultural Change, 48*, 847–865. doi:10.1086/452480

Bassi, N., & Kumar, M. D. (2011). Can sector reforms improve efficiency?: Insights from irrigation management transfer in Central India. *International Journal of Water Resources Development, 27*, 709–721.

Bhalla, G. S., & Singh, G. (2009). Economic liberalisation and Indian agriculture: A statewise analysis. *Economic and Political Weekly, 45*, 34–44.

CWC. (2005). *Water and related statistics*. New Delhi: Central Water Commission, Ministry of Water Resources, Government of India.

CWC. (2007). *Pricing of water in public system in India*. New Delhi: Central Water Commission, Ministry of Water Resources, Government of India.

CWC. (2010). *Pricing of water in public system in India*. New Delhi: Central Water Commission, Ministry of Water Resources, Government of India.

CWC. (2015). *Financial aspects of irrigation projects in India*. New Delhi: Central Water Commission, Ministry of Water Resources, Government of India.

Deshpande, R. S., & Narayanamoorthy, A. (2001). Issues before the second irrigation commission of Maharashtra. *Economic and Political Weekly, 36*, 1034–1043.

GOI. (1972). *Report of the second irrigation commission*. New Delhi: Ministry of Irrigation and Power, Government of India.

GOI. (1992). *Report of the committee on pricing of irrigation water*. New Delhi: Planning Commission, Government of India.

GOI. (2007). *Maharashtra state development report*. New Delhi: Planning Commission, Government of India.

GOM. (2003). *Maharashtra state water policy: 2003*. Mumbai: Government of Maharashtra.

GOM. (2005). *Maharashtra management of irrigation systems by farmers act (act no. XXIII of 2005)*. Mumbai: Government of Maharashtra.

GOM. (2006a). *Report on water audit of irrigation projects in Maharashtra: 2004–05*. Mumbai: Water Resources Department, Government of Maharashtra.

GOM. (2006b). *Report on benchmarking of irrigation projects in Maharashtra: 2004–05*. Mumbai: Water Resources Department, Government of Maharashtra.

Gulati, A., Meinzen-Dick, R., & Raju, K. V. (2005). *Institutional reforms in Indian irrigation*. New Delhi: Sage.

Hussain, I., & Hanjra, M. A. (2004). Irrigation and poverty alleviation: Review of the empirical evidence. *Irrigation and Drainage, 53*(1), 1–15. doi:10.1002/(ISSN)1531-0361

IWMI. (2011). *Water user's associations in the context of small holder agriculture: A systematic review of IFAD funded water user's associations in Asia*. Colombo: International Water Management Institute.

Jagadeesan, S., & Kumar, M.D. (2015). *The Sardar Sarovar Project: Assessing economic and social impacts*. New Delhi: Sage.

Kumar, M. D. (2010). *Managing water in river basins: Hydrology, economics and institutions*: New Delhi: Oxford University Press. doi:10.1093/acprof:oso/9780198065364.001.0001

Kumar, M. D., Narayanamoorthy, A., & Singh, O. P. (2009). Groundwater irrigation versus surface irrigation. *Economic and Political Weekly, 44*, 72–73.

Kumar, M. D., Scott, C. A., & Singh, O. P. (2011). Can India rise agricultural productivity while reducing groundwater and energy use. *International Journal of Water Resources Development, 29*, 557–573.

Narayanamoorthy, A. (2001). Irrigation and rural poverty nexus: A statewise analysis. *Indian Journal of Agricultural Economics, 56*, 40–56.

Narayanamoorthy, A. (2007a). Does groundwater irrigation reduce rural poverty? Evidence from Indian states. *Irrigation and Drainage, 56*, 349–362. doi:10.1002/(ISSN)1531-0361

Narayanamoorthy, A. (2007b). Tank irrigation in India: A time series analysis. *Water Policy, 9*, 193–216. doi:10.2166/wp.2006.063

Narayanamoorthy, A. (2007c). Turnaround in financial recovery in Maharashtra's irrigation sector. *Economic and Political Weekly, 42*, 2679–2685.

Planning Commission. (2013). *Twelfth five year plan: 2012–17*. New Delhi: Government of India.

Ray, I. (2005). Get the price right: Water prices and irrigation efficiency. *Economic and Political Weekly, 40*, 3659–3668.

Saleth, R. M., Namara, R., & Samad, M. (2003). Dynamics of irrigation-poverty linkages in rural India: Analytical framework and empirical analysis. *Water Policy, 5*, 459–473.

Sodal, S. V. (2007). *Reforms initiatives in water resources sector in Maharashtra State*. Mumbai: Maharashtra Water Resources Regulatory Authority. (Retrieved from www.ielrc.org/activities/workshop_0704/content/d0724.pdf).

Vaidyanathan, A. (2013). *Water resources of India*. New Delhi: Oxford University Press.

Vermillion, D. L. (1997). *Impacts of irrigation management transfer: A review of the evidence*. Research Report 11. Colombo: International Irrigation Management Institute.

World Bank. (2002). *INDIA, Maharashtra: Reorienting government to facilitate growth and reduce poverty*. Vol. I and II, (Report No. 25053-IN). Washington, DC: Poverty Reduction and Economic Management Unit, South Asia Region, The World Bank.

World Bank. (2005). *India's water economy: Bracing for a turbulent future*. (Report No. 34750-IN). Washington, DC: Agriculture and Rural Development Unit, South Asia Region, The World Bank.

Appendix. Capital expenditures, working expenditures and gross receipts of major and medium irrigation projects in India, 1974–75 to 2013–14 (INR millions)

Year	Cumulative capital outlay	Gross receipts (GR)	Working expenses (WE)	Profit (GR – WE)	Recovery (%)
1974–75	38,478	607	946	−339	64.0
1975–76	44,543	869	954	−85	91.0
1976–77	51,378	1,047	1,128	−81	92.9
1977–78	59,963	969	1,272	−303	76.2
1978–79	69,664	1,081	1,552	−471	69.6
1979–80	80,900	1,007	1,405	−398	71.7
1980–81	93,467	1,034	2,257	−1,223	45.8
1981–82	107,905	1,202	2,653	−1,451	45.3
1982–83	123,423	1,171	2,377	−1,206	49.3
1983–84	140,421	1,651	2,739	−1,088	60.3
1984–85	159,297	1,297	3,340	−2,043	38.8
1985–86	179,712	2,238	4,869	−2,631	46.0
1986–87	202,343	1,667	4,896	−3,229	34.1
1987–88	223,112	1,387	14,003	−12,616	9.9
1988–89	246,080	1,664	21,280	−19,616	7.8
1989–90	270,888	2,076	22,238	−20,162	9.3
1990–91	305,691	2,289	24,763	−22,474	9.2
1991–92	336,888	2,273	28,033	−25,760	8.1
1992–93	371,051	3,202	31,169	−27,967	10.3
1993–94	410,804	4,774	36,299	−31,525	13.2
1994–95	458,856	4,451	43,524	−39,073	10.2
1995–96	513,469	4,982	48,190	−43,208	10.3
1996–97	568,407	4,584	54,461	−49,877	8.4
1997–98	639,842	3,633	62,577	−58,944	5.8
1998–99	710,703	4,558	72,153	−67,595	6.3
1999–2000	789,528	4,569	79,802	−75,233	5.7
2000–01	862,487	7,535	87,624	−80,089	8.6
2001–02	938,982	6,522	82,497	−75,975	7.9
2002–03	960,079	7,834	88,459	−80,625	8.9
2003–04	1,104,727	10,476	62,936	−52,460	16.6
2004–05	1,284,447	12,642	70,183	−57,542	18.0
2005–06	1,504,096	11,947	82,161	−70,214	14.5
2006–07	1,689,798	15,047	96,044	−80,997	15.7
2007–08	1,998,616	20,449	118,989	−98,540	17.2
2008–09	2,360,921	19,040	121,969	−102,929	15.6
2009–10	2,681,643	23,511	149,209	−125,698	15.8
2010–11	3,004,641	25,975	173,636	−147,661	15.0
2011–12	3,343,591	38,929	187,201	−148,272	20.8
2012–13	3,709,083	31,283	213,489	−182,206	14.7
2013–14	4,051,647	43,364	218,531	−175,167	19.8

Sources: CWC (2005, 2007, 2010, 2015).

Solarizing groundwater irrigation in India: a growing debate

Nitin Bassi

ABSTRACT
India is on a path to reduce its carbon emission intensity with a major thrust on increasing the grid-connected solar photovoltaic capacity. However, the carbon footprint in agriculture is on the rise. Heavy subsidies for electricity and diesel to pump groundwater for irrigated agriculture, combined with lack of regulations on water withdrawal, are resulting in both groundwater over-exploitation and increased carbon emissions. Some researchers and practitioners have suggested large-scale promotion of solar pumps for well irrigation as a way to make agricultural growth carbon-neutral and groundwater use in farming sustainable. This article examines whether solar pumps for groundwater irrigation are technically feasible and economically viable in India.

Overview

Globally, there has been a significant increase in the uptake of clean energy technologies in recent times. In 2015, renewable energy generation (including solar and wind) grew by an estimated 5% and accounted for nearly 23% of overall electricity generation globally. Solar photovoltaic (PV) capacity alone grew by an estimated 31% in 2015 but still accounts for only about 1% of all electricity generation. Much of the impetus for adoption of clean energy technologies is a manifestation of policies driven by concerns of energy security, prevention of local pollution and increasing climate benefits (International Energy Agency, 2016).

India, too, has embarked on a mission for large-scale adoption of clean energy technologies, including those for improvement of energy efficiency across various sectors which use conventional sources of energy. Under its Intended Nationally Determined Contributions, as a commitment to post-2020 climate action in the UN Framework Convention on Climate Change Conference of the Parties in Paris (December 2015), India has pledged to reduce the carbon intensity of its GDP by the year 2030 by 33–35% from the 2005 level. To achieve this objective, India plans to adopt clean energy, promote energy efficiency in various sectors, take steps to achieve lower emission intensity in the automobile and transport sector, put a major thrust on non-fossil-based electricity generation and promote development based on energy conservation (Government of India, 2015a).

On the clean energy front, India has set an ambitious target to increase grid-connected renewable energy production capacity from 36 GW to 175 GW by 2022. Among other renewables, it is planned that wind and biomass installed capacity will be doubled to 60 GW and 10 GW respectively, and solar capacity increased 25-fold, to 100 GW, by 2022 (Government of India, 2015a).

Considering these targets, the major push is to increase the installed capacity of solar PV-based power. This is expected to be achieved through development of solar parks, ultra-mega solar power projects and canal-top solar projects. Adoption of rooftop solar PV systems and solar PV-based irrigation pumps is also promoted through subsidies.

Major claims are made regarding the promise of solar irrigation pumps in addressing groundwater-energy nexus issues in India (Kishore, Shah, & Tewari, 2014; Shah, Durga, & Verma, 2015). These include reliable energy supply, reduced fuel imports, power-sector viability, groundwater conservation and emissions reduction. However, in comparison to large-scale solar installations which feed power into the grid, mini or small solar PV systems (such as solar irrigation pumps) are inefficient and expensive (Hernandez et al., 2014). Further, since the initial capital cost of a solar pump is about 10 times that of a conventional diesel pump, capital subsidy and financing support are required for large-scale solar-pump adoption (KPMG, 2014). More importantly, the impact on the efficiency and sustainability of groundwater use in irrigation of solar-energy-driven pumps over pumps run on subsidized electricity has never been studied empirically.

The major objective of this article is to examine the technical feasibility and economic viability of introducing solar irrigation pumps and their role in addressing problems of groundwater exploitation and energy-use inefficiency in agriculture. The rest of the article is in four sections. The next section provides an account of the potential of solar power systems and their uptake in India. The third section provides an overview of the growth in groundwater irrigation in India and summarizes its impact on water and energy use in agriculture. The fourth section offers an analysis of the climate benefits (green quotient), technical feasibility and economic viability of solar irrigation pumps, and discusses to what extent these can address the groundwater-energy nexus. It also discusses the feasibility of net metering. The concluding section discusses policy implications.

Solar power in India: potential and uptake

The maximum solar power that can be generated under optimum conditions (i.e. clear sky, sun rays falling perpendicular to the solar panel and the longest day of the year), according to estimates prepared by the National Institute of Solar Energy (NISE), is about 749 gigawatts peak (GWp). The estimates of state-wise solar power potential, made on the basis of the wasteland and urban-rooftop areas which can be used for solar PV systems (overall about 15,000 km^2), and a solar PV module efficiency of 15% (NISE, 2014), show clear interstate variation in solar power generation potential. The state of Rajasthan, in Western India, has the highest potential (142 GWp), and Goa has the lowest (0.9 GWp). The potential is 22% lower for the second-rank state of Jammu and Kashmir, and 55% lower for the third-ranked state, as compared to Rajasthan (Bassi, 2016; NISE, 2014).

There is remarkable variation in the solar power generation potential across regions as well, with the highest potential in Western India, followed by Northern, Southern, Central and North-Eastern India, and the lowest in Eastern India (Figure 1). In terms of proportional contribution to the respective region's solar power potential, Rajasthan, the state with the highest solar power potential, has 58% of the total potential of the Western Region; Madhya

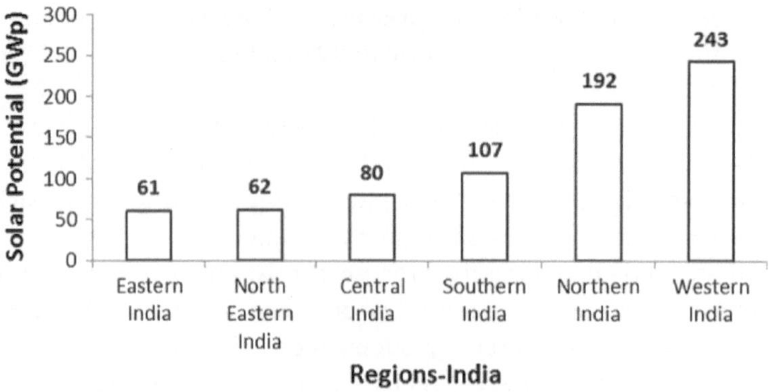

Figure 1. Region-wise distribution of solar power potential in India. Source: Based on data presented in Bassi, 2016, and in Government of India, 2014a.

Pradesh, 77% of the potential in the Central Region; Odisha, 42% of the potential in the Eastern Region; Jammu and Kashmir, 58% of the potential in the Northern Region; Assam, 22% of the potential in the North-Eastern region; and Andhra Pradesh, 36% of the potential in the Southern Region.

Compared to this huge theoretical potential, utilization has been very low, with an installed capacity of only 9.1 GW of grid-connected solar power as of 2016, constituting only 3% of the total energy mix of the country (Central Electricity Authority, 2016). A significant proportion of this capacity was installed in just the last two years (2014–16), with the government aggressively promoting it by offering a capital subsidy of up to 90%.

However, the uptake of solar PV systems is limited by numerous economic, social and policy-related constraints (Bassi, 2016). The major ones are poor economic viability owing to high capital cost; less demand for the electricity produced from solar power, owing to the energy surplus situation; time lag between installation of solar PV systems and receipt of capital subsidy from the government; and the need to secure financing from multiple sources and approval from several government agencies (Sen, 2014).

Groundwater irrigation and its energy nexus

Over the last few decades, there has been unprecedented growth in groundwater use for irrigation in India. Between 1980–81 and 2007–08, the gross area irrigated by wells grew from 17.7 million ha to 50.4 million ha, an increase of 185% (IndiaStat, http://www.indiastat.com). The intensive use of groundwater for irrigation has been made possible by the growth in energized wells. The number of electric agro-pumps increased from 0.2 million in 1961 to more than 11 million by 2007, and diesel pumps increased from 0.16 million to 6.3 million in the same period (Government of India, 2010a). Also, there has been an exponential increase in electricity use in agriculture, from 4470 GWh in 1970–71 to about 160,000 GWh in 2013–14 (Government of India, 2015b; IndiaStat), mainly driven by greater demand for groundwater for irrigation.

Heavy energy subsidies for agricultural groundwater pumping, and lack of regulations on groundwater withdrawal, are threatening the viability of the power sector and groundwater sustainability in states in semi-arid and arid regions of India. By 2011, groundwater was over-exploited in nearly 16% of the total assessment units (Government of India, 2014b), and in many hard-rock regions (which comprise 70% of the area of the country), well failures were very common (Bassi, Vijayshankar, & Kumar, 2008; Kumar, 2007; Kumar, Sivamohan, & Narayanamoorthy, 2012). On the other hand, State Electricity Boards in large parts of Northern, Western and Peninsular India, where agricultural groundwater use is high, are facing huge revenue losses owing to the heavy subsidies for electricity supplied to the farm sector (Figure 2).

Regional picture and present policy discourse

Figure 3 presents the region-wise status of groundwater development in India. The annual groundwater draft as a fraction of the net annual replenishable groundwater resources is very high in Northern and Western India. In fact, in many states of Western and Northern India (Haryana, Punjab and Rajasthan), groundwater is over-exploited, and this has resulted in groundwater scarcity and declining quality (especially salinity). Also, these are regions which experience higher electricity consumption for agriculture (up to 40% of total electricity sales, against the national average of 22%).

In groundwater-rich regions of Eastern India, such as Bihar and West Bengal, groundwater development is constrained by the limited availability of arable land. Also, there is less need for irrigation water per unit of land owing to high rainfall and a more temperate climate. Further, access to a reliable supply of electricity at affordable rates to pump groundwater economically (economic water scarcity) is a challenge for millions of small and marginal farmers, who depend either on expensive irrigation using diesel pumps or on water purchased from rich well owners (Kumar, Bassi, Sivamohan, & Venkatachalam, 2014; Scott & Sharma, 2009).

In the groundwater-scarce regions of Western, North-Western and Peninsular India, such as Punjab, Haryana, Rajasthan and Karnataka, there is a need for a policy which would restrict groundwater and energy usage while ensuring that the returns from farming are not adversely affected, especially for the small and marginal farmers. The restricted power supply for agriculture, available only six to eight hours a day, has not been effective in regulating

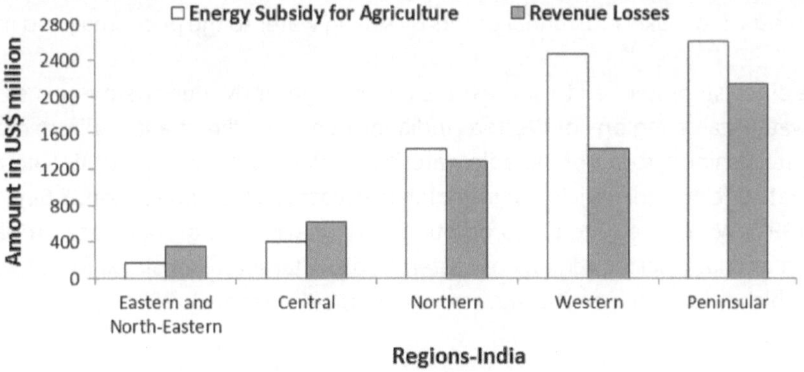

Figure 2. Region-wise energy subsidies for agriculture and revenue losses (after adjusting for subvention provided by government) for power utilities and electricity departments in India (2011–12). Source: Author's analysis based on data in Government of India Government of India, 2011.

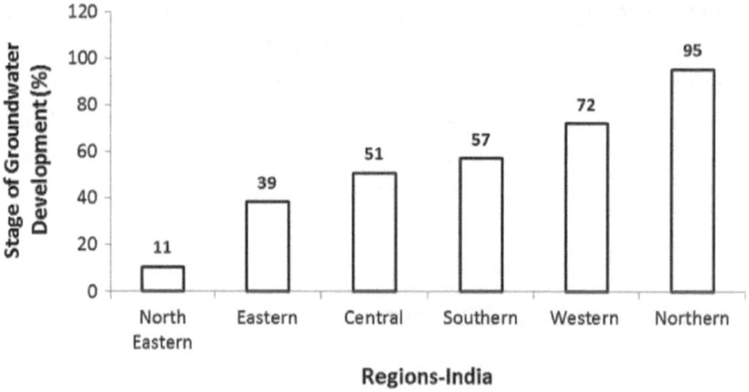

Figure 3. Region-wise stage of groundwater development in India (2011). Source: Author's analysis based on data in Government of India, 2014b.

groundwater use in this region. Adoption of solar irrigation pumps will only encourage farmers to use more groundwater, as they will be able to run their pumps even when electricity is not available from the utility, and will incur no additional cost for doing so. In contrast, in the groundwater-rich and land-scarce regions of Eastern India, which experience economic water scarcity, there is need for a policy to promote equity in access to groundwater without great cost to the exchequer (Kumar et al., 2014).

The proponents of solar PV systems for well irrigation have suggested large-scale promotion of this technology in Eastern India as a mechanism to promote equity in access to groundwater. This is based on the premise that if capital subsidy is available, the small and marginal farmers, who are now dependent on neighbouring well owners for irrigation water, could install solar PV systems in their farms and pump groundwater for irrigation instead of purchasing water from well owners at prohibitive costs, thereby drastically reducing their irrigation costs (Kishore et al., 2014). However, analysis of the impact of providing free electricity connections to farmers (in West Bengal) shows that this approach would be counterproductive, as most of the benefits of capital subsidy provision will be cornered by a few large landowners who are currently using diesel engines (Kumar, Scott, & Singh, 2013; Kumar et al., 2014). Capital subsidies for solar pumps would enable them to pump groundwater at very low costs and make substantial profits by selling water to the poor small and marginal farmers.

Some of the steps needed to address the complex groundwater-energy-nexus issues in groundwater-scarce regions of Western India include both direct and indirect measures, such as establishing systems of tradable water rights (Bassi, 2014; Kumar, 2005; Saleth, 1994) and pro-rata pricing of electricity for irrigation use (Bassi, 2014; Kumar, Scott, & Singh, 2011; Saleth, 1997). What is required to promote equity of access in groundwater irrigation in water-rich and land-scarce regions of Eastern India is targeted subsidies for micro-diesel pumps, which small and marginal farmers can adopt (Kumar et al., 2014).

The big push for solar irrigation pumps in India

In 2012, India's carbon dioxide (CO_2) emissions were about 1.97 billion tonnes, which constituted about 5.7% of the annual global emissions (Government of India, 2015a). The carbon footprint of Indian agriculture from energy use other than grid electricity (mainly diesel) is estimated at around 33.6 million tonnes of CO_2 equivalent (Government of India, 2010b). In the recent past, with growing environmental concerns about the increase in carbon footprint of agriculture, non-conventional energy, particularly solar, has attracted great attention from policy makers. The recent drop in price of solar panels worldwide, and the diesel price shocks, also prompted practitioners and policy makers to look at solar power as an important source of energy in remote rural areas, otherwise not covered under rural electrification, for both domestic use and the agriculture sector. The major policy interventions to promote the technology are in the form of capital subsidies for solar PV systems and provision for the state electricity utilities to purchase excess electricity generated by farmers connected to the existing power grids.

However, there are some important concerns related to the technical feasibility and economic viability of solar irrigation pumps and equity impacts of such policy measures. These concerns cannot be ignored, as the distribution of subsidy benefits available for purchasing solar PV systems across the landholding classes will be highly inequitable, and its overall impact on the equity of access to groundwater will be negative. Also, it could put significant financial burden on the public exchequer. Further, the claims about solar energy being 'clean' are highly questionable, as discussed below.

Are solar PV-based irrigation pumps really green?

While there is no denying that solar PV-based irrigation pumps are 'carbon neutral' in their operation, one important factor which is often overlooked in examining the clean energy benefit of the technology is the amount of energy used in the manufacturing of solar PV panels and the resultant CO_2 and other greenhouse-gas emissions – i.e. the solar PV life cycle. Solar PV panels are made of crystalline silicon, which is extracted from quartz sand at high temperatures. This process accounts for nearly 60% of the total energy requirement of PV module production. According to available estimates, the carbon footprint of PV panels over the full life cycle (mainly during extraction, transport and processing) varies from 35 g to 88 g CO_2-eq/kWh, depending on the amount of available sunlight and operational efficiency of the solar PV system (Parliamentary Office of Science & Technology [POST], 2006, 2011). These emissions are significantly higher than other low-carbon technologies for producing electricity, which include hydroelectric plants (2–13 g CO_2-eq/kWh), marine technologies (10–39 g CO_2-eq/kWh), geothermal power plants (15–53 g CO_2-eq/kWh), nuclear power plants (less than 26 g CO_2-eq/kWh), and wind turbines (20–38 g CO_2-eq/kWh) (POST, 2011). Clearly, hydroelectric plants emerge as the cleanest low-carbon technology in terms of their carbon footprint over the full life cycle.

Studies indicate that the payback period for clean energy produced by solar panels against the amount of energy required (in terms of CO_2 emissions) to produce them is up to four years for systems made of mono- or multi-crystalline silicon, which is the most commonly used (National Renewable Energy Laboratory [NREL], 2004; Peng, Lu, & Yang, 2013). The energy payback period depends on the type of panels, latitude, and cloud-free days (NREL, 2004). Considering the average technical lifetime of PV systems to be about 20 years, this is quite significant. Thus, in reality, solar panels are not carbon-free: they generate a real carbon footprint during their life cycle which is higher than other available low-carbon technologies such as hydroelectric plants and wind turbines.

Techno-social feasibility

One of the important factors influencing the performance of solar irrigation pumps is the availability of solar irradiance at a particular location (Chandel, Naik, & Chandel, 2015). The Western Region is among the best in India in terms of solar power generation potential, and the Eastern Region offers the lowest potential (Figure 1), in terms of decentralized commercial exploitation of energy (Ramachandra, Jain, & Krishnadas, 2011). The peak sunlight hours (hours of 1 kW/m^2 of solar radiation intensity) in most parts of Western India vary from 4.6 to 5.1 hours a day, whereas in eastern India it is about 3.5 to 4.0 hours a day (Table 1). Thus, the output of a solar-powered pumping system in the Western Region will be significantly higher than in Eastern India. However, even in the Western Region, it is quite common to have cloud cover, especially during the summer season. Thus, batteries are required to store power during times of good sunshine, for use on days of cloud cover when irrigation is actually required. This is important to ensure dependable power for irrigation. The battery requirement will be greater in Eastern India, due to the higher frequency of cloud cover in irrigation seasons. The less peak sunlight hours, the more expensive the required solar PV system, as more storage is needed to compensate for the limited exposure of the PV array to peak sunlight hours. Hence, in Eastern India solar PV systems will be more expensive and less reliable.

Further, in most parts of Eastern India, groundwater is shallow, and the majority of farmers depend on rented diesel pumps or their own low-cost diesel pumps to abstract groundwater. A 3 HP pump imported from China is available for as little as US$110, and a 5 HP pump for US$130 (Shah et al., 2009). In contrast, a solar irrigation pump costs US$6155–6925. Even after a subsidy of 75%, farmers have to pay around US$1540–1725 per unit. This is almost impossible for small and marginal farmers in Eastern India, whose crop production is characterized by poor yields and low incomes, to afford. Even if they borrow from banks or microfinance institutions, acquiring solar PV systems will put a significant burden on them to repay the loan, as they seldom generate any surplus from their small land holdings (Kumar et al., 2012). Thus, large-scale adoption of solar irrigation pumps is a distant possibility in Eastern India. The outcome of a policy of pushing the technology through heavy capital subsidy would be large and medium-size farmers replacing their old electric and diesel pumps with solar pumps at public cost.

There are also physical barriers to introducing PV-powered irrigation pumps in Eastern India, where the average size of land holdings is very small and both cropping and irrigation intensity are very high (Kumar et al., 2014). In states such as West Bengal, there is a dearth of barren and uncultivatable land which could be used for solar arrays (Ramachandra et al., 2011). Therefore, from the technical point of view too, large-scale adoption of solar pumps appears to be a remote possibility in Eastern India.

Table 1. Peak sunlight hours in different regions of India.

Regions	Average annual direct normal irradiance (kWh/m^2 per day)
North-Eastern India	3.4–3.8
Eastern India	3.5–4.0
Northern India	3.7–4.2
Central India	4.3–4.8
Southern India	4.3–4.8
Western India	4.6–5.1

Source: Author's estimates based on data presented in Bassi, 2016, and on India Solar Resources Maps (http://www.mnre.gov.in/sec/solar-assmnt.htm) developed by the US National Renewable Energy Laboratory in cooperation with Ministry of New and Renewable Energy, Government of India.

Though Western India has a good potential for solar power generation, there would be significant technical challenges in keeping solar-powered irrigation pumps in working condition in this region. The area has frequent sand storms and dusty winds, reducing the performance of solar cells (Chandel et al., 2015). System efficiency can be impaired by up to 18% if dust accumulates on the surface of solar panels (Sulaiman, Hussain, Leh, & Razali, 2011). The system can also have a short lifespan if dust is not removed regularly from the surface of the panels (Bassi, 2016). Thus, for optimal performance of PV systems in such areas, regular maintenance is required.

Economic viability

Solar irrigation pumps will definitely help reduce carbon emissions, and they have almost negligible operation and maintenance costs, but they require high initial capital investment (Bassi, 2016; Chandel et al., 2015; Odeh, Yohanis, & Norton, 2006; Sontake & Kalamkar, 2016). Rural areas have few technically trained personnel and poor supply chains for components and parts for solar PV systems, resulting in poor maintenance and slow repairs (Nathan, 2014). This situation can eventually affect the life of solar PV systems and therefore their economic viability (Bassi, 2016; Fedrizzi, Ribeiro, & Zilles, 2009). They are also less efficient than diesel engines. A 3 HP solar irrigation pump working for 8 hours a day using sunlight gives the same water output as a standard 5 HP pump working on grid power or diesel power for 2.5 hours a day (Bassi, 2016). A 3HP surface solar irrigation pump costs around US$5790, and a 5 HP surface-operated diesel pump only US$385.

To test the economic viability of solar pumps over diesel engines, the social costs (sum of private costs and negative externalities) and social benefits (sum of private benefits and positive externalities) of irrigation need to be estimated. Since the private benefit (incremental farm income for irrigators) accrued from irrigation is same for both technologies, comparative evaluation of the social costs and positive externalities is needed. Private costs include the cost of pumps (a 3 HP solar pump and a 5 HP diesel pump), including their operation and maintenance costs, with a negative externality in terms of the economic cost of carbon emissions from the diesel pump, which is avoided by solar PV systems.

For the economic analysis, several assumptions have to be made: (1) on average, a diesel pump will run for 200 days per year and will consume 0.75 litres of fuel per hour of operation; (2) average maintenance cost is about US$62 per year for a diesel pump and US$15 per year for a solar pump; (3) average diesel price is US$0.9 per litre in India; (4) one litre of diesel consumption will release 2.64 kg of CO_2, and the cost of mitigating 1000 kg of carbon emissions is US$7.5 (Jagadeesan & Kumar, 2015); and (5) most importantly, the life of the solar PV system is 20 years. (No scientific studies have been done in India to check the expected productive life of solar PV systems.)

Considering a discount rate of 10% annually, the present value of the social cost comes out to US$5925 for a solar pump and US$3430 for a diesel pump (Table 2). What is most remarkable is that the positive externality produced by the generation of clean power (US$65 per year) does not appear to be significant. Overall, as the results presented in Table 2 suggest, the solar irrigation pump is economically less viable than the diesel pump with respect to social costs and benefits. Therefore, providing subsidies for solar pumps on the order of 80% (US$4632 for a 3 HP solar irrigation pump, which is the average subsidy in India

Table 2. Economics of solar photovoltaic irrigation pumps in relation to diesel-powered irrigation pumps (all figures in US dollars).

Costs	Solar pump	Diesel pump
Capital cost	5,790 (for 3 HP pump)	385 (for 5 HP pump)
Net present maintenance cost	130	525
Net present fuel cost	0	2,945
Net present cost of carbon mitigation	0	65
Total cost	5,925	3,920

Source: Author's own analysis based on data in Bassi, 2015.

for the capital cost of the system) cannot be justified. The system appears to be beneficial to farmers only with heavy subsidies, which reduce the private costs.

Further, as some studies have indicated, the mismatch between water demand and supply potential makes solar PV water pumping systems oversized (Zhang et al., 2014). This can have a major effect on their cost effectiveness in terms of providing cheaper water than that from diesel pumping systems (Odeh et al., 2006). This is because in remote areas, solar PV systems are designed on the basis of the estimated demand for the peak demand month or on the basis of an occasional demand, where peak demand may occur on certain days or weeks (Odeh et al., 2006). However, in most of rural India, irrigation water demand falls significantly during monsoon months and to a large extent during summer months when wells run dry; the aggregate demand is the highest in winter months, when water is available in wells and the maximum area is irrigated. In the monsoon and summer months, oversized solar PV pumps will not be operating at their full capacity, which will eventually raise the unit cost of pumped water as the equivalent hydraulic energy (product of head and volume of water delivered per year) output falls.

Potential of net metering

One major issue with the current policy followed by the electricity utilities in many Indian states is that there is no marginal cost for the farmer who pumps groundwater, and this is one reason for over-abstraction of groundwater and its inefficient use (Kumar, 2005; Kumar et al., 2011). The solution lies in metering electricity, charging for it on a pro-rata basis, and rationing energy supply to the farmer (Kumar et al., 2011; Zekri, 2009). However, some researchers have argued that metering of electricity use in agriculture involves high transaction costs and that raising the power tariff is socially unviable and politically infeasible. They advocate the use of solar PV systems to circumvent this 'vexed' problem (Shah, Scott, Kishore, & Sharma, 2007; Shah, Verma, & Durga, 2014). Other researchers have pointed out that adoption of solar-powered irrigation pumps will aggravate the problem of groundwater overdraft because users are not confronted with the marginal cost of running their pumps (Bassi, 2016; Yu, Liu, Wang, & Liu, 2011). To address concerns related to the pervasive incentives for over-pumping groundwater, it is suggested that incentives be offered to farmers to make optimum use of solar pumps for pumping groundwater (Shah et al., 2014), so as to save the surplus electricity produced from the system for sale. However, for this incentive structure to work, three things are required: (1) infrastructure to connect farmers' solar PV-based pump sets to the electricity grid; (2) an attractive price for the surplus power they generate; and (3) net metering (Kishore et al., 2014). Net metering (charging for energy consumed minus a credit for energy generated and fed back to the grid) becomes an incentive system that credits solar energy system owners for the net amount of electricity they supply to the grid. The rationale is that this will create a high opportunity cost for inefficiently using the electricity generated from solar power for irrigation due to the fear of losing revenue, and hence would motivate farmers to save energy and water in agriculture. Significant additional income is also visualized for farmers from the sale of surplus solar power (Shah et al., 2014). To achieve this, a power purchase agreement by the State Electricity Boards which is highly favourable to the farmers is also envisaged. It is proposed that the surplus electricity generated from solar pump irrigation in every village be evacuated at a single point (presumably by establishing a micro- or mini-grid) which the electricity utility can measure and monitor (Shah et al., 2015), and that the utility should buy power at a rate of US$0.12–0.14 per kWh.

However, such recommendations have several flaws on technical and economic grounds. First, it is difficult to connect millions of scattered wells, fitted with solar pumps (earlier operating with diesel pumps), to the power grid. Second, establishment of even village-level micro- or mini-grids would involve significant financial costs (Bassi, 2016). For instance, the installation of a 250 kW solar-powered mini-grid to supply power to two villages in Uttar Pradesh cost about US$0.9 million, or about US$3845 per kW (Siddiqui, 2015). There would be an additional cost for transmission or distribution of the surplus electricity from such village-level mini-grids to other locations. Third, the utility supplies electricity to farmers at an average price of US$0.01–0.12/kWh. Hence, there will be little incentive for electricity utilities to purchase surplus electricity from the farmers at a rate 12–15 times that of the heavily subsidized electricity which they supply to the same farmers for agriculture. For instance, the state electricity board in Gujarat decided to pay only US$0.07 per kWh of electricity produced by the only solar irrigation cooperative, while the agency which promoted the cooperative demanded US$0.11–0.12 per kWh (Government of India, 2016). In 2013–14, energy subsidies to agriculture cost almost US$10.5 billion, and the electricity utilities incurred a loss of US$5 billion (Government of India, 2014c). Under such circumstances it will be financially risky for power utilities to extend any power purchase guarantee to farmers, which will produce no social benefits.

It does not make any sense to first invest in such an economically unviable technology for producing electricity and then make additional investments for costly infrastructure to buy the same electricity, in the name of incentives for conservation. On the other hand, much less investment is required to meter electricity consumption by agro-wells, paving the way for charging users of farm wells on a pro-rata basis. The 'transaction cost of metering', which is used as convenient argument against pro-rata pricing of electricity (Shah et al., 2007), can be greatly minimized with pre-paid electronic meters which work through scratch cards, or with remotely sensed meters (Zekri, 2009). And the levels of pricing at which the demand for electricity and groundwater becomes elastic with respect to tariff are socio-economically viable (Kumar, 2005; Kumar et al., 2011).

Conclusions and policy implications

To avert an impending energy crisis resulting from a rapidly widening gap between energy demand and supplies, India will have to look for different sources of energy, both conventional and non-conventional. In the wake of the growing concerns about increasing carbon emissions into the atmosphere, non-conventional renewable energy sources appear quite appealing, though policy uncertainties, non-economic barriers and grid integration continue to be major challenges to their large-scale uptake (International Energy Agency, 2016). In this article, a detailed analysis has been presented on the technical feasibility, economic viability and equity aspects of solar PV systems. The analysis shows that the technology does not generate significant welfare gains to justify heavy capital subsidies to help farmers completely switch over to solar pumps from diesel and electric pumps, whose use is already the result of heavy subsidies defended by various interest groups. Ideally, government subsidy of any technology or production system is preferred when the private benefits of the system do not offset the full costs, but the social benefits far exceed the social costs, and with the subsidy, the private costs to the adopter can be lowered. But this doesn't seem to be the case for the solar PV irrigation pumps (Bassi, 2015).

However, the government policies in the energy sector, particularly with regard to provision of subsidies, are not influenced by such economic cost-benefit analyses. This is evident in the decision of the government of India to allocate US$1.5 billion to promote solar irrigation pumps in the 12th Five-Year Plan period (Pearson & Nagarajan, 2014). This is expected to replace 200,000 old diesel pumps in the country – pumps which at the present market rate would cost only US$77 million. Considering the economic analysis presented in Table 2, the net present worth of the additional cost of maintenance of 200,000 solar pumps would be US$27 million, against US$105 million for an equal number of diesel pumps. The benefit of carbon emissions reduction would be a mere US$13 million (welfare gain) over a period of 20 years, while the total saving in diesel costs would be US$585 million over 20 years – the life of the solar pumps assumed in the cost-benefit analysis. For the conditions that exist in rural India, especially in the hot and arid areas which offer high power generation potential, a 20-year life for the solar PV systems is highly optimistic. Clearly, there is no benefit to the economy, even when one considers the welfare benefit of reduced carbon emissions. Instead, there is a loss of US$785 million (after deducting the present capital, maintenance, fuel and carbon mitigation costs of the diesel pumps from the total subsidy allocation for the solar pumps).

Instead of investing heavily in solar irrigation pumps, the government should invest in installing grid-connected mega solar power projects, improving rural electrification, metering agro-wells and charging for farm electricity use on pro-rata basis in semi-arid and arid regions of India, and developing good models for subsidies for micro-diesel engines for marginal farmers in Eastern India. Solar energy, when generated on a large scale, will be competitive without any artificial boosts (KPMG Economic Research & Policy Consulting, 1999). In addition, metering of electricity consumption by agro-wells and charging users of farm wells on pro-rata basis should receive priority, especially in naturally water-scarce regions that also experience groundwater overdraft (Kumar et al., 2011). The future discourse on addressing the groundwater-energy nexus needs to be based on sound physical, technical and economic principles.

Acknowledgment

I would like to acknowledge valuable suggestions and inputs provided by Dr M. Dinesh Kumar, executive director of the Institute for Resource Analysis and Policy, on earlier versions of the article, which helped improve the quality of analysis and arguments. The contribution of Ms Meera Sahasranaman, a research consultant at the same institute, in providing technical and editorial inputs is also acknowledged.

Disclosure statement

No potential conflict of interest was reported by the author.

References

Bassi, N. (2014). Assessing potential of water rights and energy pricing in making groundwater use for irrigation sustainable in India. *Water Policy, 16*, 442–453. doi:10.2166/wp.2013.123

Bassi, N. (2015). Irrigation and energy nexus: Solar pumps are not viable. *Economic & Political Weekly, 50*, 63–66.

Bassi, N. (2016). Managing groundwater energy nexus in India: The curious case of using solar irrigation pumps with drip systems. In P. K. Viswanathan, M. D. Kumar, & A. Narayanamoorthy (Eds.), *Micro irrigation systems in India: Emergence, status and impacts* (pp. 155–167). Singapore: Springer.

Bassi, N., Vijayshankar, P. S., & Kumar, M. D. (2008). Wells and ill-fare: Impacts of well failures on cultivators in hard rock areas of Madhya Pradesh. In M. D. Kumar (Ed.), *Managing water in the face of growing scarcity, inequity and declining returns: Exploring fresh approaches* (Vol. 1, pp. 318–330). Hyderabad: International Water Management Institute.

Central Electricity Authority. (2016). *All India installed capacity (in MW) of power stations*. New Delhi: Central Electricity Authority, Ministry of Power, Government of India.

Chandel, S. S., Naik, M. N., & Chandel, R. (2015). Review of solar photovoltaic water pumping system technology for irrigation and community drinking water supplies. *Renewable and Sustainable Energy Reviews, 49*, 1084–1099. doi:10.1016/j.rser.2015.04.083

Fedrizzi, M. C., Ribeiro, F. S., & Zilles, R. (2009). Lessons from field experiences with photovoltaic pumping systems in traditional communities. *Energy for Sustainable Development, 13*, 64–70. doi:10.1016/j.esd.2009.02.002

Government of India. (2010a). *Minor irrigation census (2006–2007)*. New Delhi: Ministry of Water Resources, Government of India.

Government of India. (2010b). *India: Greenhouse gas emissions 2007*. New Delhi: Ministry of Environment and Forest, Government of India.

Government of India. (2011). *Annual report (2011–12) on the working of State power utilities & electricity departments*. New Delhi: Power & Energy Division, Planning Commission, Government of India.

Government of India. (2014a). *State wise estimated solar power potential in the country*. New Delhi: Ministry of New & Renewable Energy (Solar R&D Division), Government of India.

Government of India. (2014b). *Dynamic ground water resources of India (as on 31st March 2011)*. Faridabad: Central Ground Water Board, Ministry of Water Resources, River Development & Ganga Rejuvenation, Government of India.

Government of India. (2014c). *Annual report (2013–14) on the working of State power utilities & electricity departments*. New Delhi: Power & Energy Division, Planning Commission, Government of India.

Government of India. (2015a). *India's intended nationally determined contribution: Working towards climate justice*. New Delhi: Ministry of Environment, Forest and Climate Change, Government of India.

Government of India. (2015b). *Energy statistics 2015*. New Delhi: Central Statistics Office, Ministry of Statistics and Programme Implementation, Government of India.

Government of India. (2016). Dhundi village in Gujarat shows the way for cooperative solar enterprise. *Akshay Urja, 10*, 44–45.

Hernandez, R. R., Easter, S. B., Murphy-Mariscal, M. L., Maestre, F. T., Tavassoli, M., Allen, E. B., … Allen, M. F. (2014). Environmental impacts of utility-scale solar energy. *Renewable and Sustainable Energy Reviews, 29*, 766–779. doi:10.1016/j.rser.2013.08.041

International Energy Agency. (2016). *Tracking clean energy progress 2016: Energy technology perspectives 2016 excerpt IEA input to the clean energy ministerial*. Paris, France: Author.

Jagadeesan, S., & Kumar, M. D. (2015). *The sardar sarovar project: Assessing economic and social impacts*. New Delhi, India: Sage.

Kishore, A., Shah, T., & Tewari, N. P. (2014). Solar irrigation pumps: Farmers' experience and state policy in Rajasthan. *Economic and Political Weekly, 49*, 55–62.

KPMG. (2014). *Feasibility analysis for solar agricultural water pumps in India*. New Delhi, India: KPMG and Shakti Sustainable Energy Foundation.

KPMG Economic Research and Policy Consulting. (1999). *Solar energy: From perennial promise to competitive alternative*. Hoofddorp, The Netherlands: Author.

Kumar, M. D. (2005). Impact of electricity prices and volumetric water allocation on energy and groundwater demand management: Analysis from western India. *Energy Policy, 33*, 39–51. doi:10.1016/S0301-4215(03)00196-4

Kumar, M. D. (2007). *Groundwater management in India: Physical, institutional and policy alternatives*. New Delhi: Sage.

Kumar, M. D., Bassi, N., Sivamohan, M. V. K., & Venkatachalam, L. (2014). Breaking the agrarian impasse in eastern India. In M. D. Kumar, N. Bassi, A. Narayanamoorthy, & M. V. K. Sivamohan (Eds.), *The water, energy and food security nexus: Lessons from India for development* (pp. 143–159). Abingdon: Routledge.

Kumar, M. D., Scott, C. A., & Singh, O. P. (2011). Inducing the shift from flat-rate or free agricultural power to metered supply: Implications for groundwater depletion and power sector viability in India. *Journal of Hydrology, 409*, 382–394. doi:10.1016/j.jhydrol.2011.08.033

Kumar, M. D., Sivamohan, M. V. K., & Narayanamoorthy, A. (2012). The food security challenge of the food-land-water nexus in India. *Food Security, 4*, 539–556. doi:10.1007/s12571-012-0204-1

Kumar, M. D., Scott, C. A., & Singh, O. P. (2013). Can India raise agricultural productivity while reducing groundwater and energy use? *International Journal of Water Resources Development, 29*, 557–573.

Nathan, H. S. K. (2014). Solar energy for rural electricity in India: A misplaced emphasis. *Economic and Political Weekly, 49*, 60–67.

National Institute of Solar Energy. (2014). *State wise estimated solar power potential in the country*. New Delhi: Ministry of New and Renewable Energy (Solar R&D Division), Government of India.

National Renewable Energy Laboratory. (2004). *PV FAQs*. Washington, DC: US Department of Energy Office of Energy Efficiency and Renewable Energy.

Odeh, I., Yohanis, Y. G., & Norton, B. (2006). Economic viability of photovoltaic water pumping systems. *Solar Energy, 80*, 850–860. doi:10.1016/j.solener.2005.05.008

Parliamentary Office of Science and Technology. (2006). *Carbon footprint of electricity generation* (Number 268). London, UK: Author.

Parliamentary Office of Science and Technology. (2011). *Carbon footprint of electricity generation* (Number 383). London, UK: Author.

Pearson, N. O., & Nagarajan, G. (2014, February 7). Solar water pumps wean Indian farmers from archaic grid. *Live Mint*. Retrieved from http://www.livemint.com/Industry/z4qrahaESPaSI5IVyQnXdJ/Solar-water-pumps-wean-Indian-farmers-from-archaic-grid.html

Peng, J., Lu, L., & Yang, H. (2013). Review on life cycle assessment of energy payback and greenhouse gas emission of solar photovoltaic systems. *Renewable and Sustainable Energy Reviews, 19*, 255–274. doi:10.1016/j.rser.2012.11.035

Ramachandra, T. V., Jain, R., & Krishnadas, G. (2011). Hotspots of solar potential in India. *Renewable and Sustainable Energy Reviews, 15*, 3178–3186. doi:10.1016/j.rser.2011.04.007

Saleth, R. M. (1994). Towards a new water institution: Economics, law, and policy. *Economic and Political Weekly, 29*, A147–A155.

Saleth, R. M. (1997). Power tariff policy for groundwater regulation: Efficiency, equity and sustainability. *Artha Vijnana: Journal of The Gokhale Institute of Politics and Economics, 39*, 312–322.

Scott, C. A., & Sharma, B. (2009). Energy supply and the expansion of groundwater irrigation in the Indus-Ganges Basin. *International Journal of River Basin Management, 7*, 119–124. doi:10.1080/15715124.2009.9635374

Sen, S. (2014, May 27). Govt's lack of foresight, political controversies and budget problems lead to slow pace of PV installations. *Electronics Bazaar*. Retrieved from http://electronicsb2b.efytimes.com/govts-lack-of-foresight-political-controversies-and-budget-problems-lead-to-slow-pace-of-pv-installations/#

Shah, T., Durga, N., & Verma, S. (2015, April 1). Harvesting solar riches. *The Financial Express*. Retrieved from http://www.financialexpress.com/article/fe-columnist/harvesting-solar-riches/59262/

Shah, T., Scott, C., Kishore, A., & Sharma, A. (2007). Energy-irrigation nexus in South Asia: Improving groundwater conservation and power sector viability. In M. Giordano & K. Villholth (Eds.), *The agricultural groundwater revolution: Opportunities and threats to development* (pp. 211–242). Oxfordshire: CABI Publishing.

Shah, T., Ul Hassan, M., Khattak, M. Z., Banerjee, P. S., Singh, O. P., & Rehman, S. U. (2009). Is irrigation water free? A reality check in the Indo-Gangetic Basin. *World Development, 37*, 422–434. doi:10.1016/j.worlddev.2008.05.008

Shah, T., Verma, S., & Durga, N. (2014). Karnataka's smart, new solar pump policy for irrigation. *Economic and Political Weekly, 49*, 10–14.

Siddiqui, R. (2015). Kannauj home to UP's first 100% solar powered villages. *Times of India*. Retrieved from http://timesofindia.indiatimes.com/india/Kannauj-home-to-UPs-first-100-solar-powered-villages/articleshow/47889179.cms

Sontake, V. C., & Kalamkar, V. R. (2016, July 1). Solar photovoltaic water pumping system – A comprehensive review. *Renewable and Sustainable Energy Reviews, 59*, 1038–1067. doi:10.1016/j.rser.2016.01.021

Sulaiman, S. A., Hussain, H. H., Leh, N. S. H. N., & Razali, M. S. (2011). Effects of dust on the performance of PV panels. *World Academy of Science, Engineering and Technology, 58*, 588–593.

Yu, Y., Liu, J., Wang, H., & Liu, M. (2011). Assess the potential of solar irrigation systems for sustaining pasture lands in arid regions: A case study in Northwestern China. *Applied Energy, 88*, 3176–3182. doi:10.1016/j.apenergy.2011.02.028

Zekri, S. (2009). Controlling groundwater pumping online. *Journal of Environmental Management, 90*, 3581–3588. doi:10.1016/j.jenvman.2009.06.019

Zhang, J., Liu, J., Campana, P. E., Zhang, R., Yan, J., & Gao, X. (2014). Model of evapotranspiration and groundwater level based on photovoltaic water pumping system. *Applied Energy, 136*, 1132–1137. doi:10.1016/j.apenergy.2014.05.045

Index

Note: **Bold** page numbers refer to tables and *italic* page numbers refer to figures.